STARSCAPES
TOPICS IN ASTRONOMY

Gerrit L. Verschuur
Fiske Planetarium, University of Colorado

STARSCAPES
TOPICS IN ASTRONOMY

Little, Brown and Company
Boston   Toronto

Copyright © 1977 by Little, Brown and Company (Inc.)

All rights reserved. No part of this book may be reproduced in any form or by any electronic or mechanical means including information storage and retrieval systems without permission in writing from the publisher, except by a reviewer who may quote brief passages in a review.

Library of Congress Catalog Card No. 76-49411

First Printing

Published simultaneously in Canada by Little, Brown & Company (Canada) Limited

Printed in the United States of America

The chapters listed below originally appeared in Astronomy Magazine and are reprinted here with the publisher's permission:

Chapter 4 — July 1974, Vol. 2, No. 7, pp. 32–36. Copyright 1974 by AstroMedia Corp.
Chapter 5 — March 1976, Vol. 4, No. 3, pp. 28–33. Copyright 1976 by AstroMedia Corp.
Chapter 6 — February 1975, Vol. 3, No. 2, pp. 34–37. Copyright 1975 by AstroMedia Corp.
Chapter 7 — April 1975, Vol. 3, No. 4, pp. 34–38. Copyright 1975 by AstroMedia Corp.
Chapter 8 — April 1974, Vol. 2, No. 4, pp. 38–42. Copyright 1974 by AstroMedia Corp.
Chapter 9 — March 1974, Vol. 2, No. 3, pp. 19–23. Copyright 1974 by AstroMedia Corp.
Chapter 10 — November 1973, Vol. 1, No. 4, pp. 20–23. Copyright 1973 by AstroMedia Corp.
Chapter 11 — December 1973, Vol. 1, No. 5, pp. 27–29. Copyright 1973 by AstroMedia Corp.
Chapter 12 — March 1975, Vol. 3, No. 3, pp. 34–37. Copyright 1975 by AstroMedia Corp.
Chapter 13 — February 1976, Vol. 4, No. 2, pp. 26–30. Copyright 1976 by AstroMedia Corp.
Chapter 14 — December 1974, Vol. 2, No. 12, pp. 36–39. Copyright 1974 by AstroMedia Corp.
Chapter 15 — June 1975, Vol. 3, No. 6, pp. 6–23. Copyright 1975 by AstroMedia Corp.
Chapter 17 — January 1975, Vol. 3, No. 1, pp. 46–49. Copyright 1974 by AstroMedia Corp.
Chapter 18 — September 1974, Vol. 2, No. 9, pp. 20–26. Copyright 1974 by AstroMedia Corp.
Chapter 20 — October 1975, Vol. 3, No. 10, pp. 50–53. Copyright 1975 by AstroMedia Corp.
Chapter 21 — July 1975, Vol. 3, No. 7, pp. 34–37. Copyright 1975 by AstroMedia Corp.
Chapter 22 — January 1974, Vol. 2, No. 1, pp. 42–45. Copyright 1973 by AstroMedia Corp.

Chapter 23 — February 1974, Vol. 2, No. 2, pp. 28–31. Copyright 1974 by AstroMedia Corp.

Chapter 24 — "Faster Than Light," August 1975, Vol. 3, No. 8, pp. 26–29. Copyright 1975 by AstroMedia Corp.

Chapter 25 — May 1975, Vol. 3, No. 5, pp. 44–47. Copyright 1975 by AstroMedia Corp.

Parts of chapter 16 appeared as "Star Death" in the Griffith Observer, September 1975, pp. 2–7. Reprinted with the publisher's permission.

Photo Credits:

Bart J. Bok, Professor Emeritus of Astronomy, University of Arizona — pages 53, 63, 65, 66, 100, 101, 123.

Hale Observatories — pages 4, 10, 20, 36, 50, 51, 60, 78, 87, 88, 97, 104, 107, 109, 111, 118, 120, 121, 132, 134, 135, 136, 137, 142, 144, 145, 148, 156, 161 (photo by Harlton Arp), 162 (photo by Harlton Arp), 163, 166, 168, 175, 178, 188.

Lick Observatory — pages 40, 128.

NASA, Jet Propulsion Laboratory — pages 13, 14, 15, 16, 17.

Also:

Page 6, Courtesy of Lockheed Missiles and Space Company; page 7, High Altitude Observatory and NASA; page 9, John A. Eddy; page 28, Max Planck Institut für Radioastronomie, Bonn, Germany; page 31, Gerrit L. Verschuur; page 45, © 1974 by the American Astronomical Society. All rights reserved. The Astrophysical Journal, Supplement Series #237, Volume 27, February, 1974, "The Third Uhuru Catalogue of X-ray Sources," p. 40. Reprinted by permission of the University of Chicago Press and American Science and Engineering, Inc.; page 70, The Kitt Peak National Observatory; page 71, National Radio Astronomy Observatory; page 76, copyright by the National Geographic Society—Palomar Observatory Sky Survey. Reproduced by permission from the Hale Observatories; page 79, Graphic Film, Inc.; page 151, composite painting of the Milky Way, Lund Observatory; page 169, diagram made with Westerbork Radiotelescope and is taken from work by G. K. Miley, G. C. Perola, P. C. von der Kruit, and H. von der Loon. Published in Nature, Volume 237 (1972), p. 269.

PREFACE

These are exciting times in the field of astronomy. Since the 1950's, previously unexplored domains of the universe have been opened up by the development of new methods of research. No longer are astronomers limited to simply peering at the sky through giant telescopes. Our knowledge has been expanded by the study of radiations other than light. For example, radio signals as well as X-ray emissions from distant objects of the cosmos are now routinely being gathered by astronomers. Modern techniques, combined with traditional methods of study, are revolutionizing our view of the universe. Objects that were invisible to us are now revealed for the first time. Our discoveries show us a wondrous universe filled with fascinating and puzzling phenomena, many of which are not yet fully understood.

Because the new astronomy can only be practiced with sophisticated methods that require complex and expensive equipment, laypersons who are interested in the field have to depend on popular magazines and journals for information about recent discoveries. The pace of exploration is so rapid that new discoveries are hardly reported in the scientific journals before newer ones are upon us. Few nontechnical or nonmathematical presentations are made available; the curious nonscientist has no easy access to the scientific reports, and often the reports are too technical to understand even when they are available.

The aim of Starscapes is to provide the interested reader with an accessible, popular-level description of the new astronomy, in an easy-to-read format. It is a collection of brief, illustrated, topical essays on some of the most fascinating current findings. Although several of the essays were written especially for this book, most were originally written for and published in Astronomy Magazine, which is addressed to readers who lack an extensive technical background. The essays focus on subjects that are of interest to the general public; no previous exposure to astronomy is necessary to enjoy reading Starscapes. Each essay explains the concepts needed to understand the discoveries being discussed. The readings are divided into seven sections, each with its own introduction. A glossary has also been provided to help explain the terms that are probably not in everyone's vocabulary. This book should serve as a useful and fascinating supplement for general and topical courses in astronomy for non-specialists.

    I am very grateful to Steve Walther, the publisher of Astronomy, for his cooperation in granting permission to use the articles that originally appeared in his beautiful magazine. I am also grateful to Terence Dickinson, who worked hard on the editing of earlier versions of many of these essays. Special thanks are extended to those who reviewed the text throughout its development: Professor Clarence T. Daub, San Diego State; Dr. Stephen Hill, Michigan State; Professor Colton Tullen, County College of Morris; and Dr. Michael Zeilik II, University of New Mexico. Valuable suggestions for the improvement of the text were also made by: Professor Paul Boynton, University of Washington; Dr. Mark Chartrand, Hayden Planetarium; Professor John C. Evans, Kansas State University; Professor Andrew Fraknoi, Canada College; Dr. William Kaufmann III, University of California, Los Angeles; Dr. Henry Salz, Springfield Technical Community College; and Professor Elske V. P. Smith, University of Maryland. Finally, my thanks are due to Susan Lynds for her helpful comments on an early manuscript for Starscapes.

CONTENTS

SECTION I	**THE SOLAR SYSTEM**		1
	CHAPTER 1	The Sun and the Moon	3
	CHAPTER 2	Mars	12
	CHAPTER 3	Jupiter, Saturn, and Titan	18
SECTION II	**OTHER EYES**		25
	CHAPTER 4	What Radio Eyes Would See	27
	CHAPTER 5	Infrared Astronomy	33
	CHAPTER 6	Ultraviolet Astronomy	38
	CHAPTER 7	X-Ray Astronomy	42
SECTION III	**INTERSTELLAR SPACE**		47
	CHAPTER 8	Between the Stars	49
	CHAPTER 9	Magnetic Fields Beyond Earth	56
	CHAPTER 10	Jewels of the Universe	62
	CHAPTER 11	Inside and Behind the Orion Nebula	69
	CHAPTER 12	Interstellar Molecules	74

SECTION IV	**STARS**		**83**
	CHAPTER 13	How Far Is Up?	85
	CHAPTER 14	Measuring Star Diameters	92
	CHAPTER 15	From Dust to Dust	96

SECTION V	**THE DEATH OF STARS**		**115**
	CHAPTER 16	Star Death	117
	CHAPTER 17	Why Does the Crab Nebula Shine?	126
	CHAPTER 18	The Dumbbell, the Owl, and the Eskimo	131

SECTION VI	**GALAXIES**		**139**
	CHAPTER 19	Islands in the Sky	141
	CHAPTER 20	The Shape of the Milky Way	150
	CHAPTER 21	Hidden Mysteries of the Galactic Center	155
	CHAPTER 22	Interacting Galaxies	160
	CHAPTER 23	Exploding Galaxies	166

SECTION VII	**QUASARS AND THE UNIVERSE**		**171**
	CHAPTER 24	Quasars	173
	CHAPTER 25	Retesting Relativity	182
	CHAPTER 26	Was There a Beginning?	186

SECTION VIII LIFE IN SPACE 193
CHAPTER 27 Where Is Everybody? 195

GLOSSARY 199

STARSCAPES
TOPICS IN ASTRONOMY

SECTION 1
THE SOLAR SYSTEM

Our star, the sun, and its attendant planets form the solar system. It is one of probably billions of similar systems in the galaxy of stars that we call the Milky Way. The importance of the sun was recognized by ancient people who worshiped the sun as their most powerful god. And today, as then, the sun is recognized as the source of life-giving energy on our planet earth. Accompanying the sun in its endless path through space are the nine planets, many orbited by their own satellites. We can now view these planets up close by using remote-control television cameras. Such exploration of the solar system is telling us more about its origins and broadening our understanding of the earth and its neighbors.

CHAPTER 1
THE SUN AND THE MOON

About 200,000 million stars are in our Milky Way galaxy alone! There are billions of similar galaxies in our known universe. About 6,000 stars are visible to the naked eye, all shimmering through the earth's atmosphere. They seem constant, shining endlessly; yet they were once invisible clouds of gas and, one day, they will die and be replaced by others.

Stars are sometimes found in clusters, large and small. There are blue stars, red stars, young stars, old stars, cool stars, hot stars, and dead stars. Stars were mapped and patterned by ancient people, who created legends to explain the unexplainable; to understand Orion, Taurus, Pisces, Cygnus, and Gemini. Now we know better—or do we?

Every morning the sun, our own star, lifts over the horizon to light another day. Worshiped throughout history, the sun shines down on us, providing warmth and driving the growth of plants, animals, and humans. Often we talk in a condescending tone about the ancients worshiping the sun god Ra. But we are still sun worshipers — look around and you will see the modern celebration of the sun. Advertisements filled with sunny scenes, cheerful symbols, and happy sunbathers sprawled in the energies radiating down are all evidence that the sun is still important to us.

But what creates the heat and the light from the sun? We no longer require gods or magic to explain that, because after thousands of years, we think we know something at last. Our interest is no longer blind worship. Instead, we study eclipses and radio waves, and use telescopes to watch events occurring on the sun's surface. We watch and measure and compute, and try to understand what it is beneath the sun's surface that generates all this energy.

Imagine Cape Kennedy, Pad Number 39,

4 THE SOLAR SYSTEM

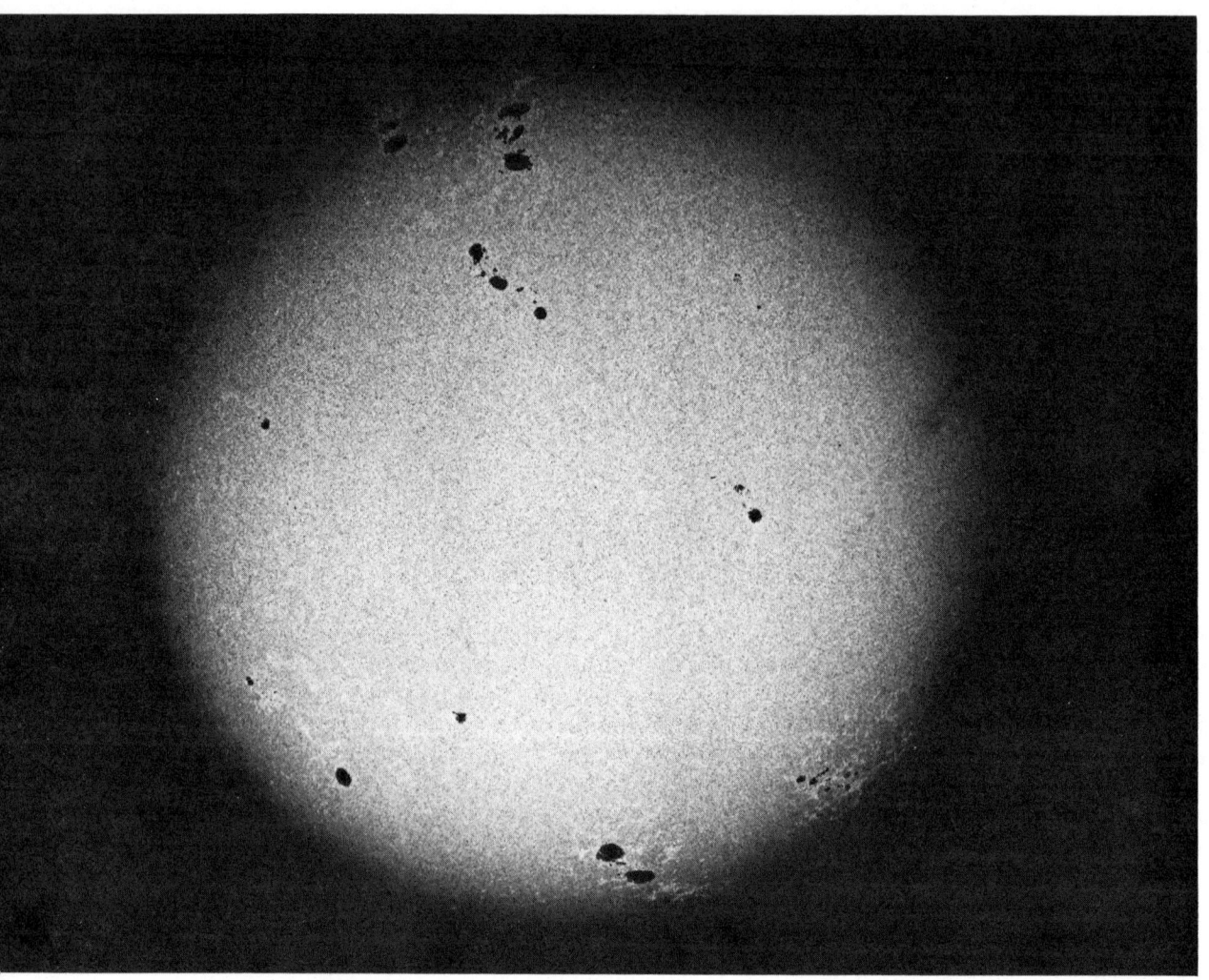

May 14, 1973. Poised to roar upward into orbit stands Skylab, a rocket manned by humans who will study our planet and our sun for months in a way never done before. Emitting a body-shaking roar, Skylab lifts off slowly, losing a solar panel and a heat shield in the process.

Human ingenuity and hard work enabled that voyage to be salvaged. A parasol sewed

The sun shows a large number of sunspots — dark blotches that are regions of somewhat lower temperatures than their surroundings.

by hand made the difference: The astronauts used the parasol to protect the damaged sections of Skylab from the sun. When they turned their telescopes on Sol, they watched events never seen before. They saw solar flares and explosions in a new light. They watched as the sun's convulsions sent clouds up and out into space, traveling outward to reach us on earth.

Deep inside the sun, enormous heat (15 million degrees Kelvin)[1] fuses hydrogen atoms together to make helium. In a million years the heat generated by the fusion filters upward to the photosphere, the visible disk of the sun, and radiates out into space. Our star is our private fusion reactor in our corner of the Milky Way. Because the sun is located 150 million kilometers away we have no waste problems or engineering difficulties from trying to contain the reaction. Our star is out there now, burning slowly, while we use its constantly available energy. It has been there for 5 billion years and will probably be there another 5 billion years before it heats up, expands, and engulfs the planet Mercury and perhaps even Venus and earth. Then it will shed an outer shell, forming a planetary nebula and the rest will collapse, fade, and die.

In the meantime, the sun is up there radiating all that energy streaming outward, adding to the light from the Milky Way galaxy that reaches other galaxies millions of light-years away. All we have to do is to collect some of that solar energy for ourselves, just as planets and living beings have been doing for billions of years. Only now we must collect it more efficiently. There is no shortage of heat reaching the earth from the sun. One day's worth of sunlight falling on the United States could supply the country's energy needs for 800 years. The ultimate energy shortage will only come 5 billion years from now, when the sun reaches the final stages in its evolution.

Humans have been watching the sun through telescopes for hundreds of years. They saw that the sun was far from being a perfect, unblemished sphere. Galileo Galilei found the blotches that freckle the sun's face, dark markings we now call sunspots. These sunspots are regions of cooler matter (4500°K) surrounded by normally hot (6000°K) matter spread throughout the visible disk of the sun. Above the photosphere, an atmosphere called the corona reaches millions of miles into space. Only during a total solar eclipse, when the visible disk is exactly covered by the moon's intrusion between the sun and us, can we see this corona shining in the skies like a beautiful halo.

The sun is not only an object visible to the eye and optical telescopes. It is also the most powerful source of radio signals in the sky, simply because it is so near. The motions of the hot gases in the solar corona cause electrons to collide and radiate radio signals that tell us about the temperatures and densities in the corona. Sometimes solar flares (explosions in the photosphere) send out clouds of particles that trigger radio signals as they move up through the corona. Such signals reach us as bursts of radio

---

[1] We will often refer to the temperature of an object. Astronomers use the Kelvin scale of temperature. A Kelvin degree (°K) is equivalent to a Centigrade degree (°C), but the zero point is different. On the Kelvin scale, absolute zero or 0°K, is equal to a temperature of −273°C. The Fahrenheit degree (°F) is generally not used in this book, but for comparison remember that 0°C, the temperature of melting ice, is 32°F and the change of one degree Centigrade is equivalent to 1.8°F change. Absolute zero is about −460°F.

6 THE SOLAR SYSTEM

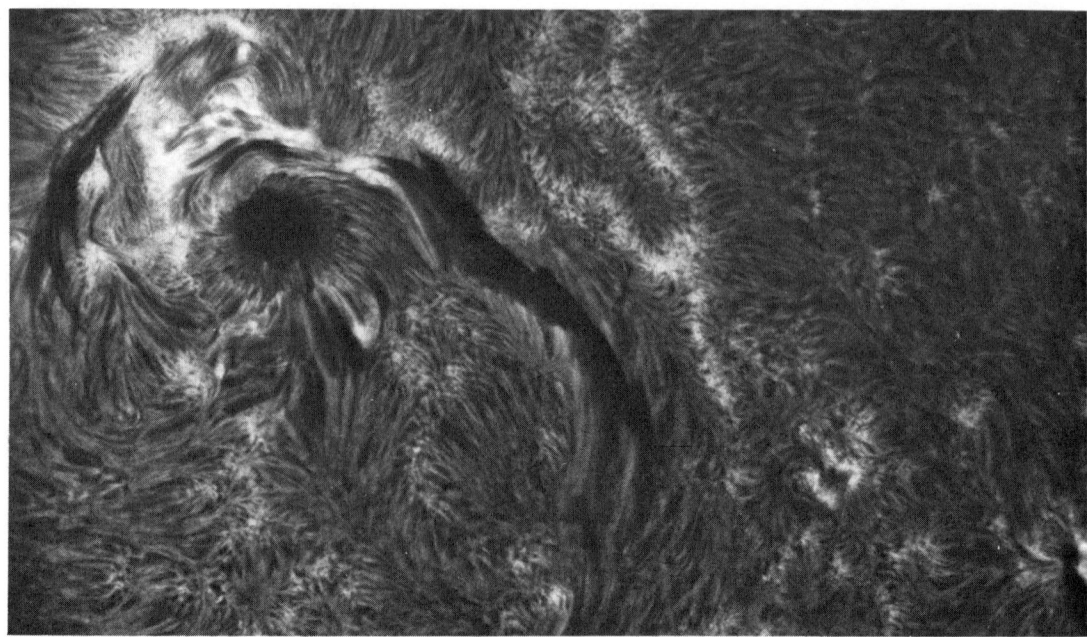

A close-up photo of a sunspot and some of the beautiful filamentary structure on the surface of the sun, photographed in the light of the hydrogen atoms on the sun.

radiation. The study of solar bursts gives us clues about the energetics of the explosions and the way matter is continually being ejected from the sun. Other forms of radiation (see section II) are also emitted by the sun.

X-ray telescopes directed at the sun reveal that X rays are emitted by the whole solar disk. The X rays come from the lowest parts of the corona and the lighter and darker structures of the X-ray image show that the sun has spots in the X-ray portion of the spectrum, too.

Ultraviolet and infrared radiations also come from the sun. Both of these are substantially absorbed by various regions in the earth's atmosphere. Water vapor cuts out much of the infrared rays, while the layer of ozone high in our atmosphere absorbs some of the ultraviolet radiation. Ozone is a molecule made up of three oxygen atoms, unlike the oxygen molecules we breathe. (The latter consist of pairs of oxygen atoms.) Only with telescopes launched above the atmosphere can the sun be studied at wavelengths usually absorbed by the atmosphere.

The Skylab astronauts made many special observations of the sun. They carried a device called a coronograph, built especially to allow them to photograph the thin corona. Observations of the corona are usually possible only during a total eclipse of the sun. Using the coronograph, the Skylab crew discovered a totally new phenomenon occurring in the corona. Large, arch-shaped structures moved outward through the corona. These arches are called transients,

THE SUN AND THE MOON 7

A solar transient photographed from Skylab on June 10, 1973. The dark disk in the center of the photo is actually in the telescope to cut out the light from the solar disk. Its diameter is one and a half times that of the solar disk and the extent of the image is six times the diameter of the sun. In time, the transient slowly moves out from the sun like a bubble being blown away.

because of their relatively brief existence (a few hours) as they moved upward from the sun and out of sight.

Before the advent of photography, astronomers relied entirely on drawings and sketches of astronomical objects for research and teaching purposes. In particular, astronomers took it upon themselves to sketch the structure of the solar corona during solar eclipses, hoping that these permanent records would help them solve the question of what caused the corona. During most solar eclipses, the corona streams out from the sun, forming a halo in the sky. On July 18, 1860, an eclipse occurred that was sketched by astronomers who had assembled along the line of the eclipse path in Spain. Notable among the drawings produced was one by Italian astronomer Guglielmo Tempel. His drawing showed a whorl-like pattern in the corona; the other astronomers sketched loop-like structures with less clarity. No such structures had ever been noted before, nor have they ever been seen again during an eclipse. As a result, those old observations were treated with skepticism and were used as proof to illustrate that little reliance could be placed on drawings made by observers, regardless of their training as astronomers. In 1900, the famous astronomer Ellery Hale said that "the experience of previous eclipses has shown that drawings of the corona for the most part serve no useful purpose unless it be to illustrate the peculiarities of the draftsman."

Time proved Hale wrong in this particular matter — the Skylab astronauts photographed no less than twenty-four events of the type sketched back in 1860. These transients, usually shaped like giant bubbles, move upward very rapidly through the solar atmosphere. They appear to occur about once every 100 hours on the average, and even at that rate we would expect to see one during an eclipse only once every hundred years. This explains why the astronomers of 1860 saw one but none had been seen since.

The transients are hard to see because they occur in the outer part of the corona where the light intensity is very low. The coronograph on board Skylab used a special filter that allowed less light in from the bright part of the corona and more light in from the dimmer, outer parts of the sun's atmosphere. Through this filter, the transients showed up clearly.

If we are lucky enough to see a transient during a future solar eclipse, it will be an incredible experience, not only because of its rarity, but because of the dramatic structures we would see in the beautiful corona.

An interesting sidelight to this story concerns Father Secchi's drawing of the 1860 eclipse. Secchi, a more renowned astronomer than Tempel, was apparently interested in the lower corona, and his drawing, made at a spot a few miles from where Tempel was sketching, does not show the transient at all. This led an astronomer to conclude in 1896 that observers spaced only a few miles apart would differ widely in their impressions of what the corona looked like. It is probably true that in the past anyone looking at both diagrams would have believed the more well-known observer and have discounted the other's work. However, time has proven that an observer's reputation is not a reliable basis for judging his or her work. Secchi was perhaps too interested in some detail of the corona to notice the transient, or possibly he noticed it and dismissed it unconsciously. We will never know.

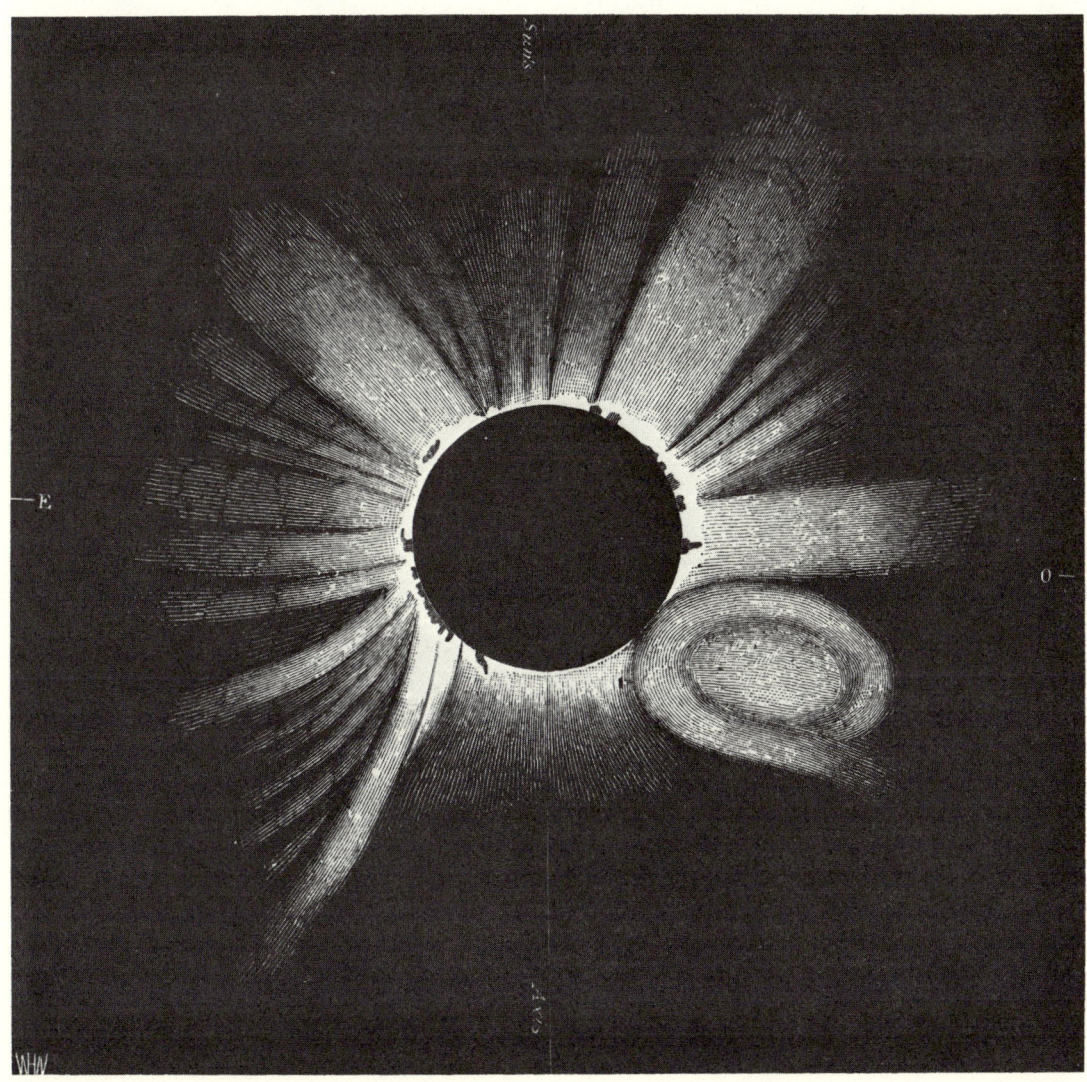

Made during an eclipse seen in 1860, this sketch of the solar corona reveals the presence of a solar transient. Only from Skylab have transients been seen again by humans.

More fascinating phenomena take place on the sun, some well understood, some hardly understood at all. The U.S. space program includes many satellites designed specifically to study the sun. We hope to learn more in the future.

The sun constantly illuminates its planets, and we depend directly upon its energy. Some of the energy is in the form of stored energy — for example, fossil fuels. Some is

The moon as seen through the 100-inch telescope of Hale Observatories. The mare (dark areas) and craters are dramatically visible in this picture.

more direct, as in the case of the solar cells that drive the spacecrafts being used to study the sun. The future story of human efforts to harness this abundant solar energy will include new and fascinating episodes that we cannot yet imagine.

The moon, too, has played an important role in the lives of people on this planet. Not only do we hear about the effects of the full moon on suicide rates, murder rates, and general moods, we also know that it is a romantic symbol in cartoons and paintings. The moon has always attracted great attention from humans, both ancient and modern. Our nearest neighbor in space, this satellite of earth has been the subject of the most extensive space explorations performed.

One basic question — how old the moon is — was answered by studying rocks found on the moon. It is about 4.7 billion years old! (It cost some $25 billion to be able to say that with conviction.) The Apollo program that put men on the moon discovered many things about this satellite, although one important question still needs to be answered: Where did the moon come from? Much was learned from the Apollo program about the evolution of the moon since its formation, but we know little of its origin. The study of lunar rock samples and seismometer recordings revealed that soon after its formation, the surface of the moon was melted by the continual bombardment of material in orbit around the sun. This process seems to have played a role for all the planets that have been closely studied. Mars and Mercury have crater-pitted surfaces just like the moon. The earth, too, must have been bombarded by ancient space debris, but erosion and continental drift have long since totally altered the structure of our planet's surface.

After the bombardment of the moon subsided, the surface cooled and became a solid rock layer, but radioactivity deep inside melted the core. Volcanic processes poured lava over the moon's surface so that the maria (the "seas" — the dark, low-lying and relatively flat areas of the moon) were covered with solidified lava (basalt). After volcanic activity ceased, the moon continued to be bombarded by meteorites from space and its surface became pitted, as we see it now. Craters of all sizes cover the moon; some have been filled in by dust created by the meteorite impacts over billions of years. The filled-in craters are the darker areas of the moon, the maria. The word implies seas, but the only thing that has flowed over those surfaces during the last 3 or 4 billion years is dust created by shattering collisions with other objects from space. Those collisions still occur, but they are few and far between, as is evidenced by the signals picked up from the seismometers left on the moon by the Apollo astronauts.

A very early suggestion about the composition of the moon was made by the scholar Erasmus in 1542, who "made his friends believe the moon to be made of green cheese." Some of the moon rock measurements suggest that this explanation is not as far off as has been believed. In experiments measuring the speed of sound in lunar samples, two scientists compared their lunar rock data with the speed of sound in many different types of earth rocks. Two lunar rocks showed sound speeds of 1.84 and 1.25 kilometers per second (km/sec), whereas typical earth rocks transmit sound at speeds of 5 to 7 km/sec. However, one class of earthly substances was found to have sound speeds of 1.5 to 1.8 km/sec — the cheeses, not all of them green. Cheddar, emmenthal, and muenster all compared well with the lunar rocks. Other differences between earthly cheeses and lunar rocks have showed conclusively that the moon is not likely to be inhabited by mice teeming in giant subterranean caverns; the velocity figures are purely coincidental.

Perhaps someday, when the manned space program regains momentum, we will have an outpost of human habitation on the moon. Our perspective has already been changed completely: The moon has been walked upon, littered with Hasselblad cameras, and marked with footprints that will remain for millions of years to come. But now, once again, it is peaceful while humans turn their attention further out into space, to Mars, Venus, Mercury, Saturn, and beyond.

CHAPTER 2
MARS

Mars is a beautiful bright planet sometimes visible in the evening — one that prompts people to phone radio stations and planetariums to ask about a UFO in the sky. Mars, the home of legends, the home of countless science fiction adventures, may someday be the home of humans.

If we stop to think about what the space program means, we cannot fail to be awed. Giant rockets have struggled off the launching pads at Cape Canaveral (or is it Kennedy again?) discarding spent sections as they surge upward and then silently swing about our home planet. Another blast sends off a smaller, instrument-loaded piece, headed for the planets.

To the planets! Twenty years ago there were educated people who did not believe we would even launch an orbiting satellite, and now we are off to Mars! The journey is unique, for on Mars we can hope to find something tremendously exciting and important — the possible existence of life. The Mariner spacecraft orbited Mars and the Viking spacecraft landed there in an effort to search for life. However, the search revealed no life.

The photographs that were sent back from the Mariner expedition are truly a feast. A whole planet is revealed for the first time. We no longer need wonder about what its surface looks like. We know that there are canyons, mountains, and dunes, dried-up riverbeds, volcanoes, and rift valleys, all on a scale that boggles the imagination.

When Mariner 9 got to Mars in 1971, a dust storm covered the planet to a depth of 60 kilometers. Mars has very little atmosphere and no water masses that could modify the storm. When the storm slowly subsided, shaded areas appeared that turned out to be dunes whose shapes changed from day to day as the winds blew.

Later photographs showed a terrain

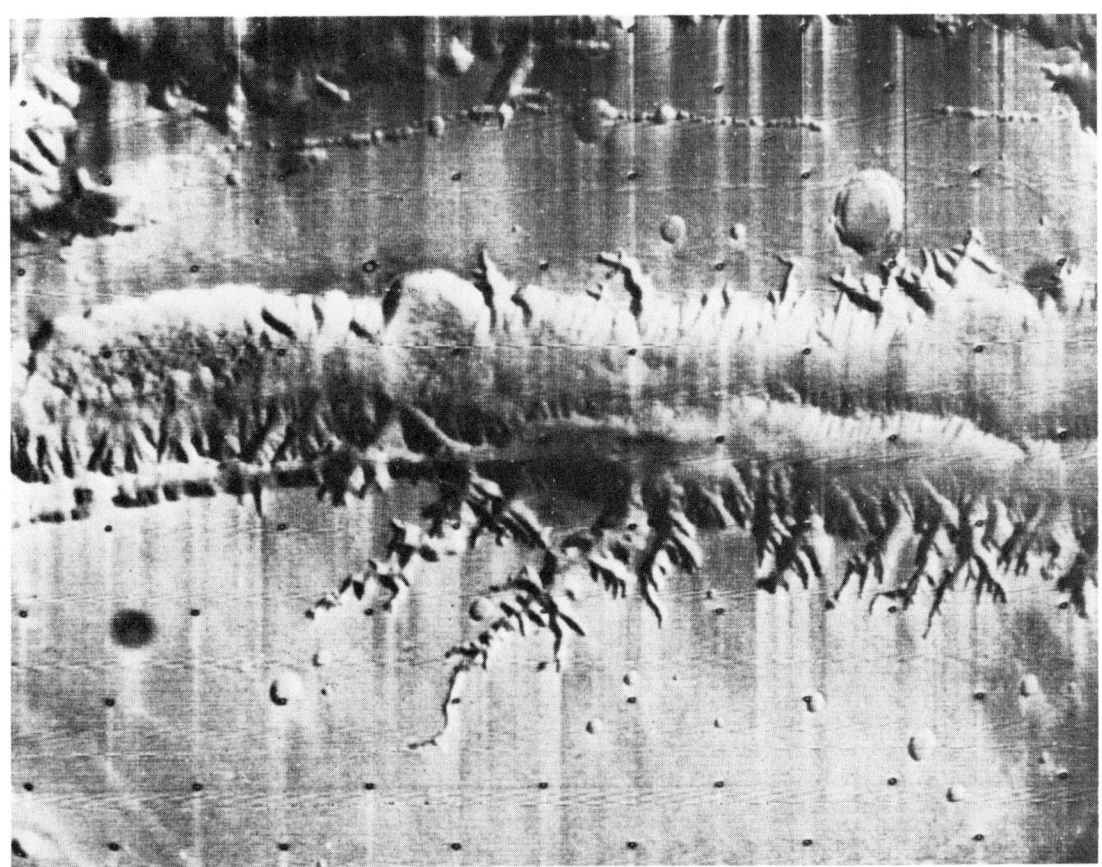

This enormous canyon on Mars has tributaries that would dwarf the Grand Canyon in Arizona. The canyon extends nearly 5,000 kilometers across Mars (the width of the United States), is hundreds of kilometers wide in places, and is nearly 10 kilometers deep. The picture covers an area of only 380 by 480 kilometers.

covered in landslides and channels. There really are channels there, but these could never have been seen from the earth as was alleged long ago. The channels head toward the east and north and do not look like lava flows. It must have been water that flowed through them. But when? And where is the water now?

There is a rift valley on Mars, 5,000 kilometers long, 8 kilometers deep, and 167 kilometers across. It is so large that a person would not even notice it if he were walking inside it.

Well-known ice caps, advancing and receding with the seasons, reveal scoured terrain under the ice worn by eons of erosion, layer upon layer of ice, and worn rock. And over the poles, clouds were seen gently drifting after the 300 kilometer per hour winds of the dust storms had died down.

14 THE SOLAR SYSTEM

This panoramic photo of Mars' horizon is the second picture ever taken from the red planet's surface. Photographed seconds after Viking 1 landed on Mars on July 20, 1976, the scene shows the Viking Lander's exterior in the foreground and a 300° view of Martian terrain. Viking's meteorology instrument is at the top left of the photo and an antenna is at the far right.

In Mare Erythraeum a dry riverbed 400 kilometers long meanders. Some still cautiously use words like "riverbed," while others talk about running water and channels. But looking at the photographs, we can try to imagine what it was like on that planet once, long ago.

Some craters have insides that look like the cracked, dried-up lake beds that one finds on earth during droughts, except on a much larger scale. And if we look at the whole map of Mars it is not hard to imagine that perhaps seas once existed. But then we leave the facts and wander into fiction. Look — there are mountains and craters, but other areas look so smooth and unstructured. Could it be that there were once bodies of water like oceans on Mars? That is not in the official interpretation (yet), but it is surely hard to imagine rivers cutting such enormous canyons as the rift valleys and their tributaries unless the water flowed somewhere to be recycled to rain down elsewhere.

The sand dune deserts on Mars catch the eye. Wind erosion was very important in their formation and is even now taking place

A channel thought to have been formed by running water in the past is seen at right in this mosaic of three pictures of Mars taken July 1, 1972, by Mariner 9. "Flow" of the channel is northward, from lower left to upper right. This small segment of the channel is about 75 kilometers in length and is located just north of the Martian equator.

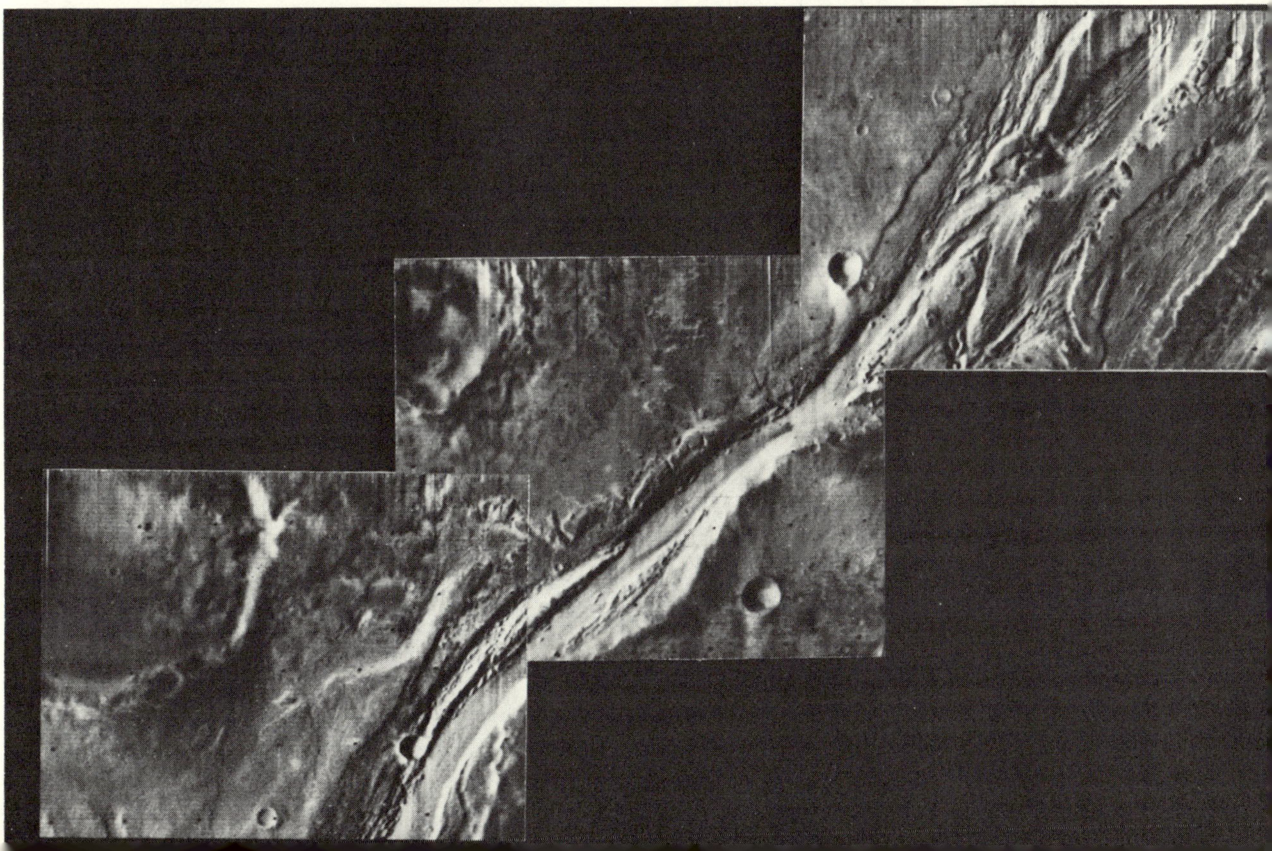

MARS 15

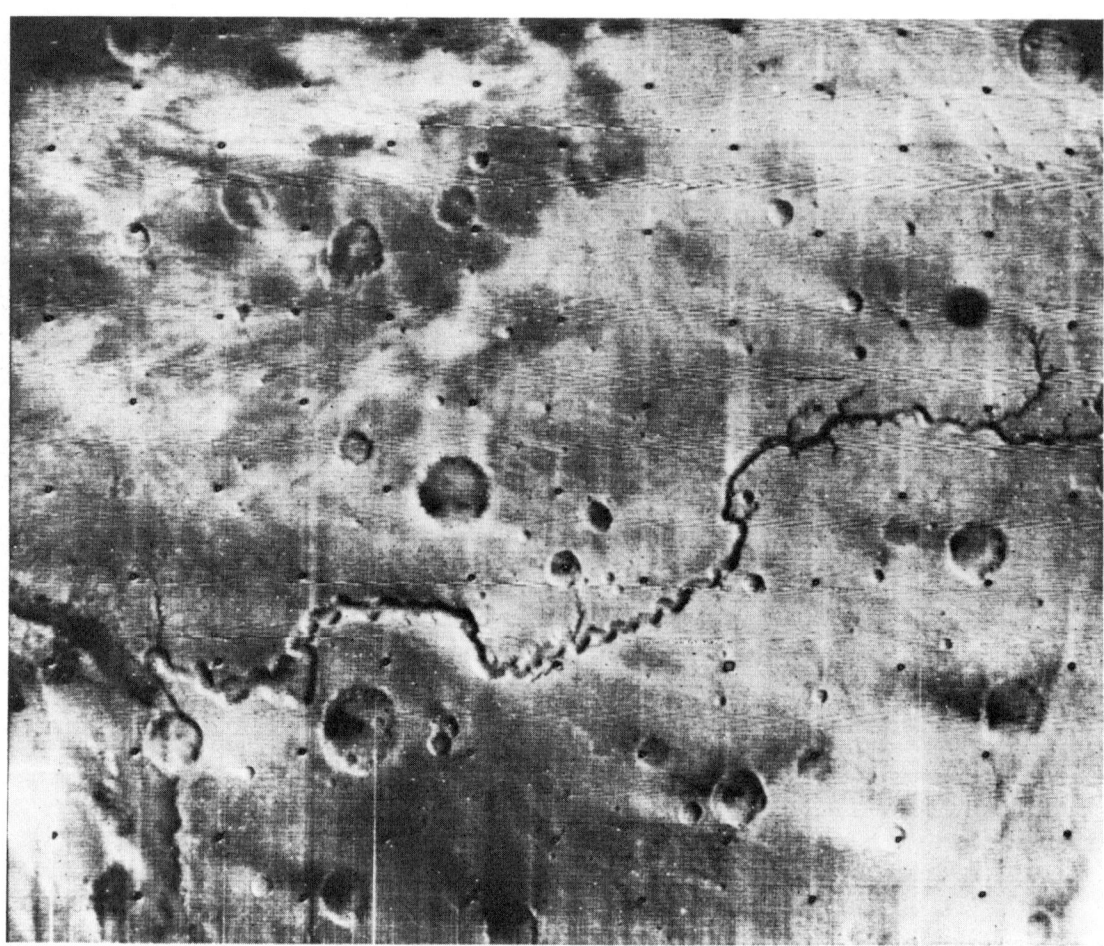

An unexpected feature on the Martian surface discovered by Mariner 9 is this sinuous valley that looks much like a river meandering its way across the landscape. Some 400 kilometers long and 5–6 kilometers wide, this valley was probably cut by flowing water; but there is now no water on the surface of Mars. What happened to this water is one of the unsolved mysteries.

— winds that sweep across the dead and dry Martian deserts and mountains. Any water that might now be at the surface will quickly evaporate into space.

But again we come back to the past. Where is the water that swept down those mountainsides and carved rivers? Is it below the surface now? Is some of it in the ice caps? Perhaps we will soon know the answer. The Viking Lander was an attempt to find more answers. Probes were extended from the spacecraft and scooped up handfuls of soil and tested them for the presence of simple living organisms to no avail. Any living things there should have been detected, provided their chemistry is like that on earth. Since we believe that Mars and earth are likely to be similar in that respect, the search was for "life as we know it."

Imagine Mars as it may have been millions of years ago: the desert-like landscape, rivers running out of the river banks, plants that survive the temperature extremes of Mars, which nowadays get to well above freezing at a point near the equator where the sun is overhead, but dip to 100 degrees Fahrenheit below freezing at night. In those days the climate must certainly have been very different. The rivers could not have been frozen and the water was not all in the form of glaciers flowing slowly down to ice seas. No, Mars must once have been warmer and must have had a substantial atmosphere. And in the presence of an atmosphere and running water did simple living things evolve at some unknown time in the past? Not just plants, but perhaps animals evolved. And were any of these creatures intelligent? Did the huge climate change that came their way come with some warning? Was it possible that a nearby supernova set off processes in the Martian atmosphere that started an irreversible change in the air of that planet that led to the total change of climate with the water being lost, the carbon dioxide ice caps forming, and the planet once again becoming exposed to the merciless bombardment of radiation from space? There is no reason to believe that any of this happened, unless one wants to sell a book on pseudoscience or science fiction, but perhaps we will have the answers some day.

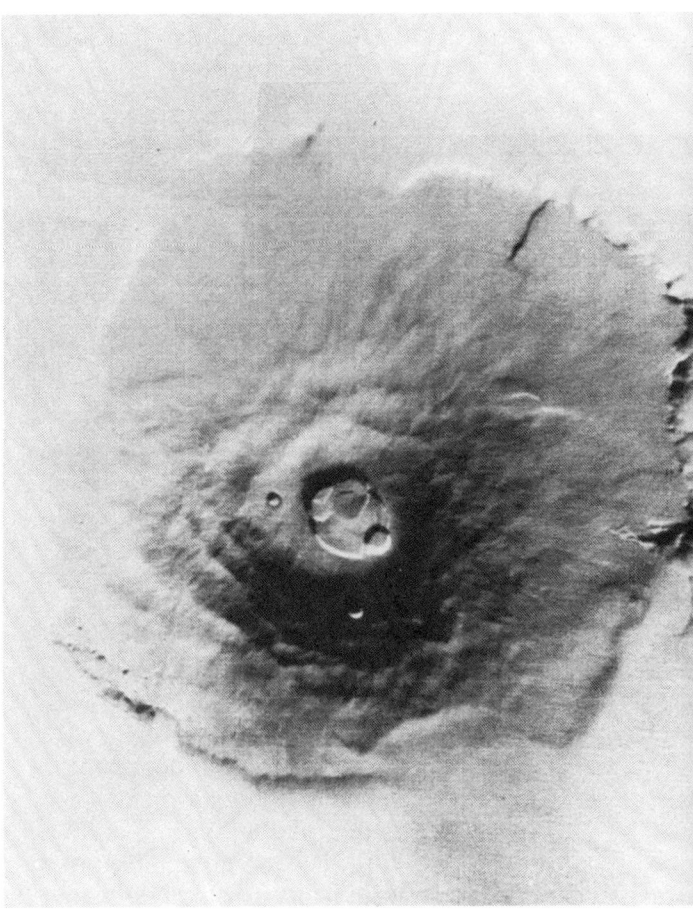

A giant volcanic mountain on Mars, called Olympus Mons, was photographed by Mariner 9 late in January, 1972. The summit stands nearly 23 kilometers above the surrounding terrain, its base is 600 kilometers across, and the main crater is no less than 65 kilometers in diameter. This mountain dwarfs anything on earth; it is nearly three times higher than Mt. Everest.

CHAPTER 3

JUPITER, SATURN, AND TITAN

There is a distant region of the solar system, so far away that it is almost a separate corner of space. The space probes sent to Mars, Venus, and Mercury have only had to travel the equivalent of the sun-earth distance, but to visit Jupiter and Saturn the spacecraft would have to travel for many years, since these two giant planets lie five and ten times as far from the sun as does the earth. These two giant planets are so different from the earth that their exploration is already one of the most fascinating tales of planetary astronomy.

In orbit around Saturn we find the largest satellite of the solar system, Titan, which is almost as big as Mars. One property that Jupiter, Saturn, and Titan have in common is that there is every reason to expect that the chemistry in some parts of their atmospheres might well enable the formation of complex organic molecules and even simple life forms. Of all the thirty-four satellites in the solar system, Titan is one of two with an atmosphere (Io, orbiting Jupiter, is the other), and its atmosphere is closer to that of earth in terms of its pressure than any of the planets' atmospheres.

Jupiter, the largest planet in our solar system, more than 300 times the mass of the earth and eleven times its diameter, is somewhere between being a planet and being a star. It did not quite make it to starhood according to recent spacecraft missions to Jupiter that have shown conclusively that Jupiter is radiating more heat than it receives from the sun. There must be a hot center to Jupiter, even though its upper cloud layers are hundreds of degrees below freezing on our Fahrenheit scale. Some estimates put the central temperature at

50,000° Kelvin, which might be caused by the slow contraction of the gaseous mass that makes up Jupiter. A decrease in Jupiter's size by only one millimeter per year could maintain the high temperature. This temperature is well below what is needed to trigger nuclear reactions that cause average stars to shine, but is much greater than anything found inside planets such as earth. Jupiter, therefore, can be thought of as a star that failed to ignite its nuclear furnace because it was not massive enough to do so.

The surface of Jupiter, as seen from outer space, is laced with parallel bands of colored clouds and marked by the famous Red Spot. We now know that the Red Spot is an enormous storm, similar to a hurricane, which has already been blowing for nearly 300 years. It covers an area as large as the surface of the whole earth! In the colored bands of Jupiter, winds of hundreds of kilometers per hour blow constantly and make chaotic eddies swirl between the borders of the band structures. The planet spins on its axis once every ten hours, which means that the matter at its equator is traveling at 34,000 kilometers per hour (compared with 1,600 km/hr for the earth).

As we look at a cross section of Jupiter's atmosphere, we find that the cloud tops are located below the top of the atmosphere that, in the highest reaches, contains mostly hydrogen and helium as well as small amounts of ethane and acetylene. The bright white clouds seen in some of the zones on Jupiter's face are made of ammonia ice crystals hanging in a mostly hydrogen and helium atmosphere 100 kilometers below the outermost levels. These clouds are similar in nature to cirrus clouds on earth. At this height the atmospheric pressure is 70 percent that of earth and a balloon could float indefinitely here; someday, balloon-borne experiments might be done high in Jupiter's atmosphere.

About 20 kilometers below the ammonia cloud tops there are brownish and orange-colored clouds thought to contain ammonia hydrosulphide crystals. Their different colors are probably due to different chemicals in the clouds.

As one plunges deeper into Jupiter's atmosphere it gets warmer, so we assume there must be regions with earth-like temperatures and somewhat higher pressures. Sinking down to 800 kilometers below the cloud tops, the temperature reaches thousands of degrees Kelvin.

In these upper parts of the atmosphere below the clouds, conditions are probably suitable for the formation of organic molecules like those found on earth that are so basic to life as we know it. This is particularly so around 100 kilometers below the cloud tops, where water droplets or snowflakes swirl about in the clouds. Lightning flashes there might provide the necessary energy allowing water, ammonia, methane, and nitrogen to combine to form life-giving molecules.

Since the planet is surrounded by belts of lethal radiation, humans will probably never visit Jupiter or its inner satellites. The presence of these belts can be inferred from the emission of strong radio signals by Jupiter that emanate from a volume of space about three times the diameter of the planet itself. These can only be explained if Jupiter has a strong magnetic field that can trap energetic particles coming from the sun. Similar, but much less intense, belts of charged particles are located about 8,000 to 50,000 kilo-

20 THE SOLAR SYSTEM

Jupiter, the giant planet, laced with dark and light bands formed by chemical coloration of the trade wind zones in Jupiter's atmosphere.

meters above the earth's surface, and here they are known as the Van Allen belts. Jupiter has a magnetic field about ten to twenty times stronger than that of the earth; its field can trap many more particles and hold them in a doughnut-shaped region significantly above the planet's surface. The particles involved are mostly electrons and protons with a few nuclei of some heavier atoms.

These charged particles are traveling around the magnetic field lines at nearly the speed of light. If they collided with living matter, they would destroy cells. The earth's belts of radiation do not contain enough high-speed particles to pose a health hazard to astronauts flying through them, but anyone venturing into Jupiter's radiation belts would receive 100 times the lethal dose of radiation. Three of the four inner satellites of Jupiter lie inside these belts and, therefore, humans will probably never visit their surfaces. There appears to be no easy way to shield people from such intense radiation.

The satellite of Jupiter known as Io (one of the Galilean moons) constantly moves through these radiation belts and appears to trigger the leakage of particles (by an unknown process) into Jupiter's atmosphere, where they probably cause dramatic aurorae. (Aurorae are called the northern lights when seen in the sky near the earth's north pole.) However, these in-falling particles also seem to generate strong radio signals (at about 15 meters in wavelength) that are easily picked up at the earth, thus making Jupiter the strongest radio source in the sky at this wavelength. As these particles fall, they emit sharp bursts of radiation, rather than constant signals, and the structure in the bursts indicates a rotation period of 9 hours 55 minutes for the source. It is believed that this rotation period applies to the interior solid regions of the planet from which the magnetic field originates. Since this rotation period is a few minutes faster than that of the visible bands on Jupiter's surface, we can assume that the atmosphere is constantly circulating over the more solid center at about 500 km/hr.

Equivalent bursts of radio signals, linked to aurorae, appear to be produced by the earth. The aurorae occur when some of the particles trapped in the earth's Van Allen belts leak down into our atmosphere. Recent observations from an Imp earth satellite show that very strong radio signals at the long wavelength of 600 meters are being generated in the auroral regions of earth. These radio signals are also being emitted in the form of bursts, probably because the in-falling particles are falling along the magnetic fields in separate clumps. The reason the particles leak down in the first place is far from being understood. To an observer outside the solar system, the earth is the strongest radio source in the solar system at the long wavelength of 600 meters. This wavelength lies in our regular broadcast band; were it not for the highly effective shield produced by the earth's ionosphere (a layer above the atmosphere that both prevents signals at this frequency from reaching us from space and causes those produced on earth to be reflected back down again), we would be swamped by the signals produced in the auroral regions. In the absence of the ionosphere, they would make normal radio reception in the broadcast band impossible.

The Imp satellite also discovered similar bursts of radio noise from the planet Saturn. Whereas Jupiter radiates strongly at a wavelength of 15 meters and the earth at 600 meters wavelength, Saturn appears as the strongest radio source in the sky at a wavelength of 300 meters. Since this wavelength is also in our usual broadcast band and is reflected by the ionosphere, we must use satellites to study these signals from above the ionosphere. The radio bursts from Saturn seem to come in a very regular way, repeating every 10 hours 30 minutes, which is nearly the same as the rotational period of that planet. The mechanism for the production of Saturn's bursts of radio signals may be similar to that found on Jupiter and the earth, implying that dramatic aurorae are visible on Saturn. Combined with the sight of Saturn's rings overhead, the aurorae must create quite a panorama viewed from Saturn.

The rings of Saturn are one of the most uncommon features in the solar system. They are flattened, about 300,000 kilometers across, and only 5 kilometers thick. Exploring these rings by spacecraft poses an interesting problem — flying along them would be hazardous because the chances of collision with particles or rocks in the rings would be high. Flying across them would only allow a much too brief glimpse of what it was we just flew through! The rings are thought to consist of particles and small rocks that might be a few centimeters across. Based on radar signals bounced off them, they appear to be rough and ice covered.

The origin of the rings is not understood. Clearly, they occur so close to the planet itself that if a moon of Saturn had been located there at some time in the past, it would have been shattered by gravitational tidal effects from the planet.

The shattering of a satellite of Saturn in this way presupposes that the moon once existed at a vulnerable distance from Saturn. However, it could not have been formed at this range and it is unlikely that it moved in from some greater distance and then shattered. Recent calculations have shown that the rings could not have lasted for as long as the planet has been there, which rules against the likelihood that the rings formed their present configuration at the same time the planet was born. Apparently, some 3 to 4 billion years ago, Saturn did not have rings. Where did they come from? Perhaps a collision between a moon of Saturn and large meteorites shattered the moon and filled nearby space with the debris, forming the rings.

Jupiter, Saturn, and Titan, the largest satellite of Saturn, have several things in common: One is that they all appear to be reddish. It also seems that the atmospheres of these three heavenly bodies might well be suitable for the formation of organic molecules and simple life forms. (Organic molecules contain carbon as an essential element.) The reddish color may be due to the presence of such organic molecules. Red is produced because the atmospheres of these objects absorb the ultraviolet and blue light that falls on them, but reflect the rest of the light.

Titan has an atmosphere, consisting mainly of methane and molecular hydrogen, whose pressure at the surface is about one-tenth that of the earth. Because of the structural changes seen in the light reflected from Titan, we know that clouds cover most of the satellite. Recent measurements of the detailed properties of this light are consistent with a nearly total cloud cover.

New measurements of the amount of infrared (or heat) radiation from Titan show that it is hotter than would be expected from the amount of sunlight falling on it, assuming that the heat can freely radiate into space again. This could be explained if there is a greenhouse effect (that is, heat being trapped under cloud layers) occurring on Titan, much in the way it does on Venus.

The temperature of Venus below its thick cloud layer is 700°Kelvin, enough to melt lead. (The greenhouse effect on earth keeps the average earth temperature above freezing.) On Titan the clouds must be transparent to visible light, but opaque to heat radiation, so that the atmosphere there is much warmer than expected.

It has been suggested that the coloration of Titan's clouds might be due to the presence of hydrocarbons being produced there. Another interesting aspect of this satellite, which is almost the size of Mars, is that there is a lot of molecular hydrogen present, and according to simple expectations, this hydrogen should rapidly escape into space. Its presence, therefore, means that it is constantly being replenished in some way. The only obvious source appears to be volcanoes that must be bubbling liquid methane, ammonia, and perhaps even water. If we allow for the existence of internal heat in Titan, then some of these volcanoes, and possibly geysers, may be hot enough so that simple organic molecules can be synthesized around them. These suggestions about Titan are still very controversial, however.

Titan will be visited by spacecraft in the 1980's and, we hope, will reveal many more secrets. With an atmosphere allowing relatively easy parachute-assisted descent to the surface, Titan is potentially one of the most interesting objects for future space exploration.

SECTION II
OTHER EYES

Light is but one form of the radiations called electromagnetic waves. We will not dwell on the physical nature of this sort of radiation, since our understanding of the discoveries of astronomy do not require an understanding of all of physics. The difference between light and other electromagnetic waves is only a difference in wavelength. Wavelength is the distance between peaks (or troughs) in the pattern that the waves set up in space about us. Other forms of these electromagnetic waves are radio, infrared, ultraviolet, X rays, and gamma rays. These waves can all be generated by humans in various ways and are also being generated by various types of objects in space. This range of waves makes up the electromagnetic spectrum. Some objects give off only radio waves, others shine just by the light they emit, and yet others send out only X rays. Special telescopes are used to detect all of the different electromagnetic waves found in the universe. Seen through these instruments, the sky looks different at each wavelength region of the spectrum.

Wavelengths of these radiations range from one billionth of a centimeter for gamma rays, to a few hundred-thousandths of a centimeter for light to a thousandth of a centimeter for infrared. Radio waves span the range of a few millimeters to many hundreds of meters in length. There are basic technological and physical differences in the way these radiations are detected. For example, photographic plates are useless for "seeing" radio waves. We have to build sensitive radio receivers and antennas to discover radio signals from space.

Light covers only an incredibly small one ten-billionth of the range of wavelengths that reach us from space. Discoveries made by astronomers at wavelengths other than light will be discussed in this section.

CHAPTER 4
WHAT RADIO EYES WOULD SEE

What would the sky look like if we had "radio" eyes? If instead of seeing what we call visible light, we could tune our eyes to detect the radio radiation that floods through space — what would the universe look like?

Orion would be gone as would Cygnus, Sagittarius, and Scorpius — all the familiar star constellations would be invisible. A new vista of eerie radio sources would be strung in a celestial necklace across the sky near the zone where the Milky Way once was. Beyond that band, other radio sources would dot a darkened cosmos.

But what are radio sources anyway, and how are they located if our eyes cannot detect the sky's radio signals?

In the early 1930's Karl Jansky of the Bell Telephone Labs constructed a crude radio receiver that he used to detect the source of interference hampering transatlantic telephone links. He picked up a hiss that appeared to be coming from the Milky Way. A few years later, Grote Reber, an amateur radio operator, confirmed this discovery and also found that the strongest hiss came from the direction of Sagittarius behind which lies the center of our galaxy.

Twenty years later astronomers finally had a good theory for the origin of this radio hiss. It is produced by cosmic ray particles — spiraling about the magnetic fields that thread their way through all of our Milky Way. When a radio telescope capable of detecting these weak radio signals is pointed at the Milky Way, it picks up the maximum amount of this radio emission. When it is pointed away from the Milky Way, very little emission is received.

By scanning the sky systematically with a giant radio telescope, the radio astronomer can make a map of the direction from which radio signals of different intensities are being received. Such maps are drawn in the

28 OTHER EYES

The 100-meter diameter radio telescope is in West Germany, near Bonn. The inner part of the telescope consists of solid metal plates, and the outer part is made of wire mesh. Radio signals are bounced off the dish to a smaller metal mirror mounted on the tripod and reflected to the antenna mounted in the structure at the center of the main reflector. From there, the signals are fed into amplifiers and down to a control room where they are processed further.

form of contours indicating the strength of the received signals coming to us from outer space. Contours of increasing signal strength would look just like contours indicating different elevations on a topographical map. Therefore, when examining a map of the radio sky, you can identify the presence of strong sources of radio waves from space because they appear as "mountains" on the map.

The earliest radio maps of the sky showed a broad belt of radio emission closely following the outline of the hazy band of light we call the Milky Way. The highest radio contour levels on these maps are found in the direction of the constellation Sagittarius. But beyond the limits of the Milky Way the maps showed many small "hills" that seem to be spread randomly over the sky. These are the so-called radio sources. Photographs of these areas reveal some "hills" to be well-known objects, such as supernova remnants (e.g., the Crab nebula, see chapter 16) or peculiar galaxies. But at the position of many other radio sources just a faint star is seen. After the discovery of quasars in 1963, some of the radio sources turned out not to be stars at all, but quasars masquerading as stars. However, that did not explain all of them, for even now many radio sources are known to exist where absolutely nothing is visible on a photographic plate.

Radio signals are generated by different physical processes in different objects in the universe. While one object may be a strong emitter at a short radio wavelength, another might be a strong emitter at a long radio wavelength. (Stars do the same thing in visible light; hot stars emit much of their energy in short wavelengths and appear blue to the eye and cool stars, emitting energy at longer wavelengths, appear red.) The result is that the radio sky looks quite different at different radio wavelengths.

If we imagine that we have radio eyes and that a strong radio signal is the same as a bright light source, then we would see the radio sky as a whole lot of "stars" in the sky with a bright band of them where we normally see the Milky Way. We will see this at a radio wavelength of, say, 75 centimeters, or cm (one of the radio astronomers' favorite wavelengths). If we could shift to 40 cm wavelength, we would see basically the same set of stars but their relative brightnesses would have changed. In addition, the band of the Milky Way would have become narrower. Instead of being about 20 degrees wide (as it was at 75 cm), it would be about 10 degrees wide. At 20 cm wavelength, we would see that the Milky Way was even thinner — only 3 or 4 degrees wide. At all these wavelengths the Milky Way in Sagittarius would be the brightest, while toward Taurus it would be much less obvious.

If we tuned the wavelength all the way down to 2 cm, we would find the Milky Way hardly visible at all. Instead, we would see many very bright spots all along the Milky Way (or at least along the line where the Milky Way would have been). These bright spots are some of the nebulae and supernovae that are still clearly "visible" but the emission from the Milky Way itself (the emission from the cosmic rays spiraling around interstellar magnetic field lines) has become very faint.

And what would we see if we tuned our radio eyes to a wavelength longer than

75 cm? First, the width of the Milky Way would balloon until the whole sky filled with emission. The emission would still be strongest along the Milky Way band itself, but now — even at the galactic poles — there would be brightening of the whole sky. A typical wavelength at which this might be seen is about 150 cm.

If we were to look at the radio sky at an even longer wavelength something strange would start to happen. The Milky Way would become darker and darker until, at a wavelength of 30 meters, the sky would be bright everywhere except in patches along the Milky Way. There we would see regions of varying darkness, much like the dark patches produced in the visible portion of the spectrum by dust clouds. A few starlike points of light would be seen spread over the rest of the sky, but these would be mostly washed out by the sky's brightness. At a 100 meter wavelength, the Milky Way would be a dark band stretched across the sky.

These dark patches are produced by clouds of low energy, or thermal, electrons, different from the high-speed electrons that make up the cosmic rays. These clouds absorb the radio signals produced by cosmic ray electrons. At very long wavelengths we also find that some of this distant emission nebulae, visible as bright spots in the sky at short wavelengths, become dark holes absorbing all the radiation originating beyond them. To someone with radio eyes, then, the sky at long wavelengths would always be bright and there would be no day and night! The sun, of course, would shine at all wavelengths and would continue to dominate the sky.

The constellation Cygnus completely loses its identity when seen in "radio eyes." The familiar visible stars (faint gray) are invisible in this 20 cm wavelength view. An entirely new group of radio stars emerges where the "northern cross" is seen by our eyes.

Probably the most comprehensive surveys of the sky so far have been made at a wavelength of 20 cm. Let's see how the sources of radio emission other than the glow from the Milky Way are distributed over the sky at this wavelength. On page 31 we have plotted the location of the brightest sources, found at 20 cm. The symbols used were chosen by the following criteria.

In order to draw a parallel with the sky that we actually see at night, we have attributed to the radio sources a "luminos-

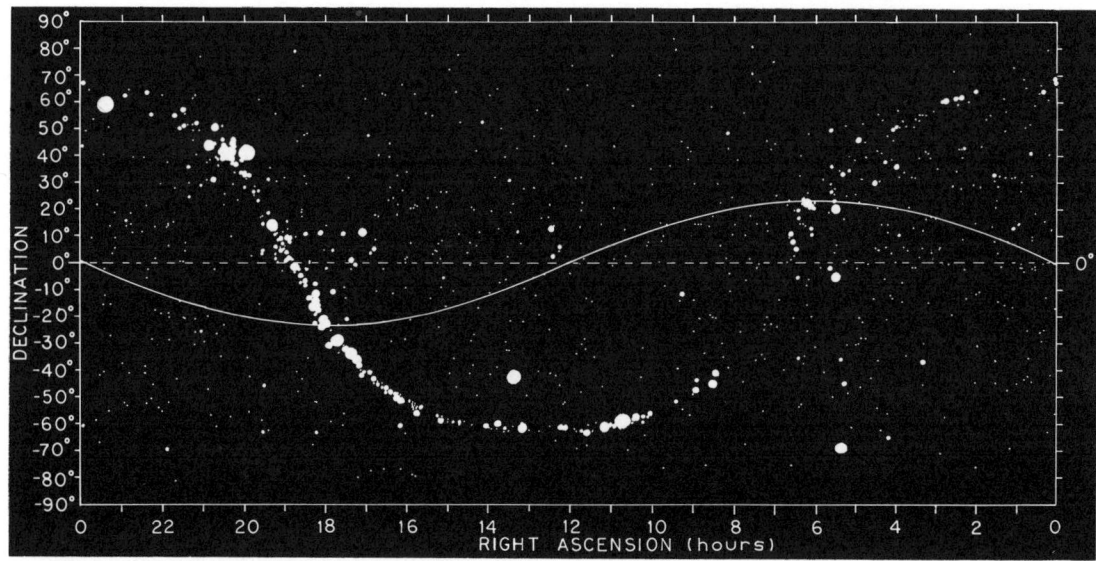

A map of the sky shows the location of the strongest radio sources as measured at a wavelength of 20 centimeters. The solid line is the path of the sun through the sky during the year. The strength of the radio emission from any given source is indicated by the size of the dot used. The strongest radio sources lie along a curved line in this map, which is the Milky Way. Most of these sources are emission nebulae or the remains of old supernovae (fainter sources are mostly galaxies or quasars).

ity" scale that mimics the optical system. The brightest visible star, Sirius, has a luminosity or magnitude of about −2. Fainter stars have magnitudes that are indicated by positive numbers. We assigned the brightest radio source, the supernova remnant Cassiopeia-A, a radio magnitude of −2. This is arbitrary since radio astronomers measure the brightness of radio sources in terms of "flux units" or Janskys, as they are now called (a measure of how much energy reaches a square centimeter of area per second). A slightly different size circle was drawn for sources of successively lesser magnitudes and the map was cut off at magnitude 6 (the symbol used for magnitudes 5 and 6 is the same). The distribution of radio sources — "radio stars" if you like — is very different from the optical sky.

Most of the bright radio sources lie in the Milky Way band, and most of the weaker sources lie over the rest of the sky. This is because the brighter sources are the nearby emission nebulae and supernova remnants in our galaxy, and the fainter sources are the distant galaxies and quasars. With the unaided eye we can see only one galaxy (in Andromeda) and no quasars, so the visible sky bears no resemblance to the radio sky. In addition, normal stars do not emit significant amounts of radio signals, so none of the familiar stars in our sky are recognizable in the radio sky. The only "star" visible to the naked eye that is also seen in the radio sky shown in our diagram is the Orion nebula (chapter 11). We have, of course, left the sun and planets out of the picture.

The great differences between the radio sky and our real sky are due to the different emission processes involved in generating light and radio waves, and the fact that many of the distant emission nebulae that might just have been visible to our eyes are obscured by interstellar dust.

The faintest radio sources so far catalogued have a magnitude of about 13 on our scale. This compares with the optical astronomers' magnitude limits of 22. Part of the reason for this apparently small range is that not many radio telescope systems are capable of detecting sources much weaker than that because of receiver sensitivity and telescope size limitations.

Strictly speaking, our magnitude scale is not comparable to the optical system at all since the optical magnitudes refer to the total visible light received on earth, and we have only talked about one radio wavelength here. If we made a map at a different wavelength, the sky would look different — not only as far as the Milky Way is concerned, but also in the way the relative brightness of the weak radio sources would change. Different "constellations" would be found at different wavelengths.

To make a better comparison with the optical magnitudes we would have to know how the radio source intensities vary over all radio wavelengths (spectrum) and then combine these to get a single number referring to the total luminosity of the sources. No one has yet done this for more than a few radio sources and certainly no one has attempted to make a map of the radio sky showing true luminosity of, say, a thousand sources. It's a gigantic task for anyone who wants to try.

CHAPTER 5
INFRARED ASTRONOMY

We can feel infrared radiation. Apart from light, it is the only form of electromagnetic radiation to which the human body is directly sensitive. But infrared radiation consists only partly of rays that we can feel as heat; some of these radiations are at wavelengths we cannot sense. The infrared region of the spectrum is located between visible light and the very short radio wavelengths. In practical units the wavelengths of infrared radiation can range from one ten-thousandth of a centimeter (called one micron) to one millimeter in length. Waves somewhat longer than one millimeter are regarded as radio waves.

The techniques for detecting waves of different wavelengths from space depend on the specific wavelength. Radio receivers — albeit very special radio receivers — are used for waves longer than one millimeter. Light-gathering systems, including photographic plates, mirrors, and lenses, are used to detect waves in the wavelength region below one micron — the visible region of the spectrum. Infrared radiation, however, does not significantly affect photographic film except at wavelengths very close to those of light. Infrared also cannot be detected by radio means. An infrared receiver uses a heat-sensitive device that allows more electrical current to flow through it in the presence of heat radiation. Such changes can be measured with a device called a bolometer; hence, infrared waves from space can be detected. A bolometer used with a mirror forms an infrared telescope. However, this telescope can only pick up infrared radiation from one small area of the sky at a time.

To study the infrared sky, astronomers have to systematically scan the sky, slowly looking at one spot after another through their telescopes. This is a very tedious process. The technique is quite similar to that used in making surveys of the sky at radio wavelengths. An additional difficulty faces infrared astronomers — infrared is absorbed by the water vapor in the earth's atmosphere. For most experiments, telescopes must be above as much of the atmosphere as possible. Early infrared telescopes were flown to high altitudes in balloons, which only allowed brief looks into space. Later, high-flying jet planes were especially equipped with infrared telescopes that could be poked through the side of the planes to get a good look at distant space. Nowadays, infrared telescopes are flown aboard rockets to get an unobscured view of the universe.

The earliest infrared experiments concentrated on looking at objects in the solar system. Since some infrared radiation does leak through the atmosphere at specific wavelengths called "windows," these telescopes were usually modifications of optical telescopes located on high mountains. By measuring the infrared emission from a planet, we can find the temperature of the planet directly. The infrared temperature refers to the temperature of the visible surface of the planet, whether it is the solid surface of Mars or Mercury, or the cloud tops of Venus and the major planets like Jupiter.

Recent infrared astronomy experiments have revealed that there are many objects outside the solar system that emit infrared more strongly than any normal star. The list of such objects includes very old stars, very young and possibly unborn stars, galaxies, and quasars. The most important infrared sources are the young stars, or protostars, since here we witness the stages in the birth of a star just before it can be said to be a true star. At this very young stage, a star does not yet generate its own heat by nuclear processes.

There are over 6,000 known sources of strong infrared emission in the sky, most of them objects in the Milky Way itself. The basic types of infrared emitters are either stars located behind a lot of interstellar dust (fine, solid matter between the stars) or stars surrounded by dust that has been heated by the stars. In the first case, a perfectly normal but very bright star might be located behind so much dust that the visible light from the star is absorbed by the dust particles. However, since these particles are not good absorbers of infrared radiation, the infrared leaks out through the dust cloud. Although we might think we are seeing an unusual infrared object, we are only seeing a normal star through lots of dust.

The second basic, and most interesting, type of infrared star is surrounded by its own dust cloud from which it has just formed. The dust that wasn't used in forming the star may well be used later to make planets. In the meantime the star heats the dust around it; the dust is reradiating this heat as infrared radiation detectable at earth, many hundreds or thousands of light-years[1] away.

These dust shells do not always originate in the leftover material from star formation. The dust shells around some stars may have been ejected from the stars themselves and then heated by the stellar radiation. Again,

---

[1] A light-year is the basic measure of distance in astronomy. It is the distance light travels in one year, about 10 trillion kilometers.

we see it as an infrared source. In this case, too, the shell absorbs the visible and ultraviolet light from the star so that we cannot see the star itself. Some of these dust envelopes or shells are very large, as much as a thirtieth of a light-year across, and the dust temperature is only 100 degrees Kelvin.

Many different types of stars are found in the infrared source list. Included are Mira variables (variable stars whose small brightness variations follow a pattern first noticed in the star Mira) and novae (stars that suddenly throw off material, brightening by millions of times as they do so), both of which appear to have associated dust shells. A class of stars known as carbon stars are also bright infrared emitters. The very youngest known stars, the T-Tauri stars, are also known to have shells of matter around them and these shells may well have resulted from gas and dust thrown out by the stars soon after they started to shine brightly.

Several infrared stars have been discovered near objects such as the Orion nebula. These infrared stars show temperatures of only a few hundred degrees and are probably protostars. Protostars are hot because they are still shrinking, and it is this shrinkage that generates heat.

The most interesting types of infrared sources in the galaxy are those associated with radio emissions from molecules of OH, a combination of oxygen and hydrogen. These OH sources emit very strong radio signals at approximately 18 cm wavelength and the strength of the signals can only be explained if the OH molecules themselves are bathed in lots of infrared radiation. The molecules absorb energy from the infrared waves and they then reradiate that extra energy as radio signals. This process is referred to as maser amplification. This OH is present in the dust shells around some of the protostars and there is now evidence for the presence of other molecules in these dust shells as well.

The center of our galaxy is also a strong and interesting emitter of infrared radiation. Around the central core, about 1 degree across in size, there appears to be infrared emission that can best be understood in terms of an enormous number of stars located behind a lot of dust that cuts out virtually all the light. A similar infrared source is found in our neighboring Andromeda galaxy. Associated with the central radio source of our galaxy, called Sagittarius A, there appears to be an infrared emitter that is probably the same object. A larger region around the galactic center, some 4 degrees long and 2 degrees wide, also emits infrared and this is probably being produced by reradiation of heat from dust around the stars in the central parts of the Galaxy.

The photograph shown on page 36 is a totally new view of the center of our Milky Way, a photograph taken at infrared wavelengths. We already mentioned that infrared detectors are quite unlike optical ones, so the use of the word photograph is very loose. To obtain this image the infrared telescope was systematically scanned in a grid pattern over the area around the galactic center and all the data were recorded. This is similar to a television picture that is produced by a scanning electron beam inside a television tube. When the data were all recorded they were processed in a computer and fed onto a television screen, where the picture was

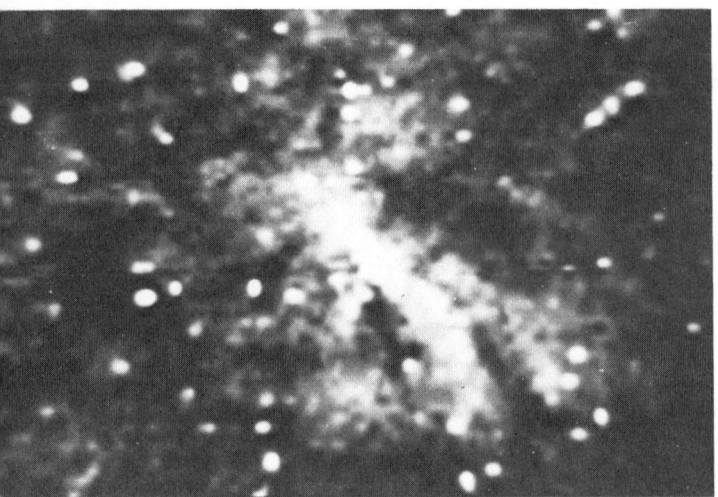

A map of the galactic center region made at infrared wavelengths (top) compared with an optical photo of the same area of sky (bottom). Since the infrared radiation penetrates through the dust clouds between us and the center of the Galaxy, while the light is totally stopped, the top photo shows relatively cool stars actually near the center of our galaxy. The bottom photo shows stars near the sun (within a few thousand light-years).

then photographed. Thus we have a photo of the infrared radiation from the center of our galaxy. The lower part of the photograph is an optical picture of the same area of sky.

The infrared photo looks so different from the optical picture because we are seeing right to the center of the galaxy, some 30,000 light-years away, but the stars in the lower picture are located only up to a few thousand light-years from us. Notice that there are many infrared stars visible as well as dark areas on the infrared picture showing that there must be large dust clouds that are so dense they also absorb most of the infrared radiation from behind them! The slight fuzziness in the photo is due to the fact that the resolution (or ability to see details) of the infrared telescope is not as good as that of the optical telescope. The infrared picture was made at a wavelength of 2.2 microns.

Other galaxies also emit infrared emission in ways similar to the Milky Way, but some unusual galaxies, called Seyfert galaxies, show much more infrared radiation than usual. Very violent motions appear to occur in their central cores, due to some explosive or similar phenomena, and the strong radio and light radiations from the central parts of Seyfert galaxies are generated by cosmic ray particles spiraling around magnetic fields. The infrared radiation from the Seyfert galaxies is being produced by these cosmic rays; these sources then are very different from any of the infrared sources found in our galaxy.

There are two very unusual infrared objects near the central plane of the Milky Way itself. They are not starlike in shape or size, but extend to cover a few minutes of arc of sky. These objects are called Maffei 1 and Maffei 2 after their discoverer, an Italian astronomer who was surveying a part of the sky with an infrared telescope. We now know these two objects are galaxies that are located behind a lot of dust in our Milky Way. Although the dust prevents their light emission from reaching us, their infrared emission manages to penetrate the dust to reach us. Following the discovery of radio signals from the Maffei objects, astronomers decided that these objects are probably galaxies located 3 to 4 million light-years from us.

CHAPTER 6
ULTRAVIOLET ASTRONOMY

Thanks to blacklight posters and related psychedelia sold over the past few years, ultraviolet (UV) light has become one of the most popularized regions of the electromagnetic spectrum. Blacklight is a fitting nickname for UV light since its wavelength falls just beyond the shortest wavelength of light our eyes are sensitive to (something like a sound that is too high-pitched to hear).

Only tiny amounts of UV light ever reach earth's surface because a protective shield of ozone gas in the atmosphere's highest layers absorbs most UV energy coming from the sun. Small quantities do penetrate, however, causing sunburn. The absorbing layer of ozone has prevented astronomers from studying UV light from the sun, distant stars, or galaxies — at least until suitable telescopes could be lifted above earth's atmosphere on rockets or satellites.

The first UV astronomy experiment was conducted after World War II with primitive devices on some captured German V2 rockets. But not until the 1960's did UV observations really begin, although they were mainly confined to solar research. NASA's 1968 launching of the first orbiting astronomical observatories at last put UV telescopes above the atmosphere for more than a few minutes at a time. And a recent satellite named Copernicus has provided the most detailed knowledge to date about the spectrum's UV portion.

Through examining UV light, astronomers can learn not only about the chemical composition of a star's outer atmosphere but also about the kinds of elements making up gases floating in interstellar space. But how can studying UV light and other forms of electromagnetic energy tell us about chemicals that are countless light-years away? The atoms in interstellar clouds block out certain wavelengths of starlight heading toward earth. Atoms of each element in the cloud absorb only a specific wavelength or set of wavelengths, giving every element its own unique absorption pattern — as individual as your Social Security number. The pattern shows up as a series of dark lines overlapping a star's spectrum.

Many interstellar elements' "spectral signatures" appear only in the UV area. The first detailed discoveries include such basic atomic building blocks as oxygen, nitrogen, carbon, magnesium, silicon, phosphorus, sulphur, iron, and titanium. Among the most exciting discoveries is that of molecular hydrogen (consisting of two hydrogen atoms) — long suspected but only detectable in UV — and molecular deuterium (one hydrogen atom and one deuterium atom). Comparing amounts of these two chemicals in the direction of the star Beta Centauri has given us new insights into the creation of the universe.

Our universe, according to the theory now most popular with astronomers, was born from the explosion of a huge, superdense ball of neutrons sometimes called the "cosmic egg." In the first few chaotic instants of the "big bang," this concentrated matter disintegrated into individual neutrons, many of which further broke apart into electrons and protons. In turn, electrons and protons paired off, creating the newborn universe's first and most abundant element — hydrogen. If the temperature of the big bang was high enough, some protons forming the nuclei of hydrogen atoms fused with free-floating neutrons, transforming a percentage of hydrogen into a new and heavier element called deuterium.

By estimating the amount of hydrogen that was turned into deuterium, astronomers hope to answer one of the most nagging fundamental questions confronting modern astrophysics: How much matter exists in the universe? If the cosmic egg was very dense, then most deuterium should have taken on additional electrons and turned into helium. Conversely, the lower the egg's density, the more deuterium we should see.

Through UV observations astronomers have found 70,000 times as much hydrogen as deuterium in interstellar clouds. This ratio indicates the proportion of the very early universe. From this ratio they calculate that if you evenly spread out all the matter of the universe, you would find one hydrogen atom in every 300 cubic meters of space. These results show that there is not enough matter to gravitationally "glue" the universe together; instead, the galaxies should continue flying apart from each other indefinitely. But if the universe was 27 times more dense than the deuterium calculations indicate, it would be gravitationally closed. The gravitational tugs of every galaxy upon every other one would continually slow down the expansion rate until the galaxies stopped for an instant, and then slowly began falling back toward ultimate collision with each other.

But these calculations of the density of the universe from the hydrogen/deuterium ratio presuppose that deuterium was created only in the big bang and not by some other way. If other forces are still pumping deuterium into space, then the hydrogen/deuterium ratio is giving us a false picture of the chemical events during the cosmos' fiery birth. No one has yet come up with an alternative for making deuterium, although observations of the speeds at which very distant galaxies are receding from us suggest that the universe is really gravitationally closed — the opposite of the UV results.

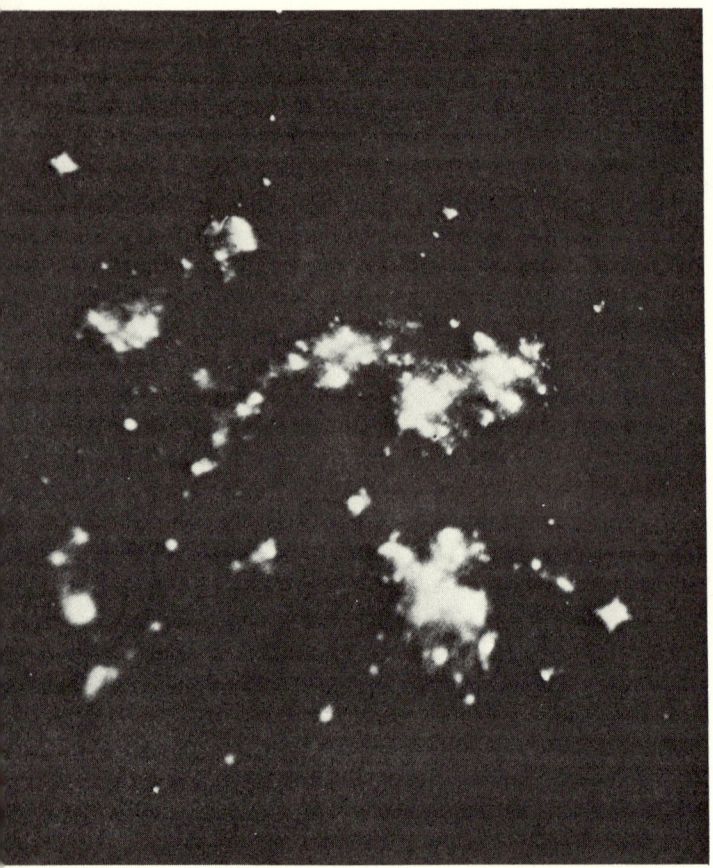 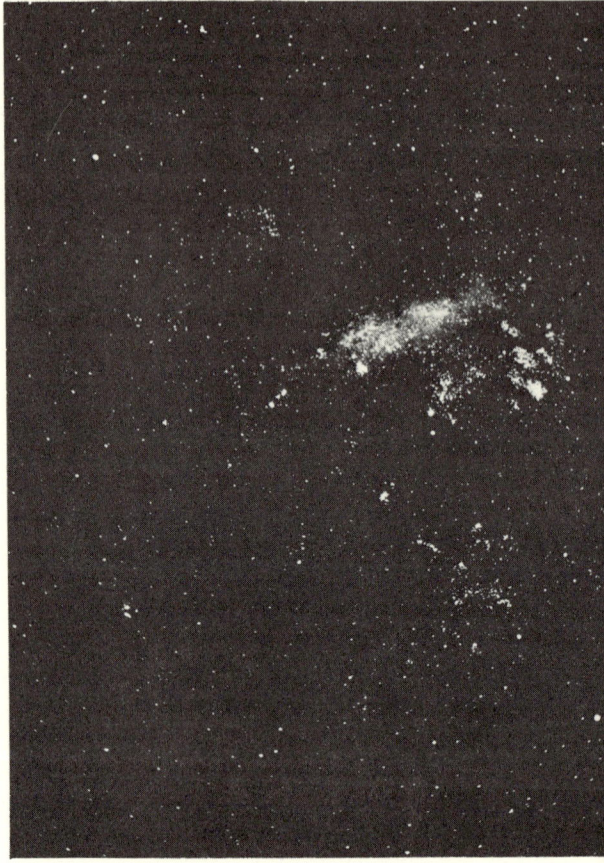

An ultraviolet photograph of the large Magellanic cloud on the left-hand side contrasted with an optical picture on the right. The ultraviolet picture reveals the presence of the hottest stars and emission nebulae in the cloud, which is a nearby companion galaxy to the Milky Way.

Besides gas, huge clouds of dust inhabiting the vast gulf between stars also absorb incoming starlight and give us clues about what kinds of chemicals make up the dust. Visible light is usually reddened as fine dust particles scatter the bluer, shorter wavelengths, and UV light gets attenuated in similar ways. Comparison of how visible and UV light are affected by their collisions with interstellar matter has revealed that some of the dust may be composed of small graphite particles.

Apparently interstellar dust clouds have distinctly different types of particles even within one cloud; in some regions of space one particle predominates, and other types of particles are more common in other regions. Some astronomers speculate that the nature of a given dust cloud depends on what types of stars have shed matter into it.

An especially exciting surprise was the discovery of interstellar clouds roasting at temperatures ranging from 100,000 to 1,000,000 degrees Kelvin. Their existence is inferred from observing characteristic UV signals from highly ionized oxygen. Astronomers previously thought that interstellar matter was much cooler, seldom rising above the 1,000 degree Kelvin mark.

Besides revealing the existence of elements in space, UV data from Copernicus cast new light on the so-called P Cygni stars — usually very hot and luminous supergiants having large quantities of matter streaming away from them. A typical P Cygni star might throw off a whole sun's mass in less than a million years. Commonly called the stellar wind, this matter hurls outward fast enough (almost 8 million km/hr) to escape back into interstellar space. Our sun also injects small quantities of matter into space, but comparing it with a P Cygni star would be like comparing a trickling faucet to Niagara Falls. UV proves to be an excellent region to clearly observe superhot atoms blowing away from the star.

Observations made on the Apollo 16 moon landing are among the most recent milestones in UV astronomy. A UV camera carried on that mission photographed earth and various star fields as well as our neighboring galaxies, the Magellanic clouds.

Pictures of earth were particularly spectacular because the hydrogen gas surrounding our planet radiates in UV and cloaks earth in a "geocorona" — much as the solar corona surrounds the sun. Being the lightest of elements, hydrogen in earth's atmosphere extends almost 100,000 kilometers into space.

By measuring UV airglow scientists learn how solar energy is distributed as it falls on the atmosphere, and also can detect small amounts of particular gases. Atoms of many gases in earth's atmosphere absorb and reradiate only certain UV wavelengths of sunlight. Just as interstellar gas has its own special signature, so do gases surveyed in our atmosphere — except that by taking a UV spectrum in this case we would find bright emission lines in an otherwise dark spectrum.

CHAPTER 7
X-RAY ASTRONOMY

Gone are the days when a dedicated astronomer sat for hours with eyes glued to the telescope, sketching craters on the moon, cataloguing new star clusters, or drawing nebulosities at unknown distances. No longer can one person spend years at one telescope and singlehandedly reveal wonders never before realized.

Today teams of scientists work together, designing and building sophisticated instruments. Rockets, satellites, and balloon-borne devices carry telescopes of shapes and functions never dreamed about twenty years ago. These telescopes are flown high above the atmosphere to probe the universe in regions where man's senses are totally useless. Up there, infrared and ultraviolet radiations, X rays, and gamma rays are streaming through space. The focus of current research is to pinpoint the sources of this exotic radiation.

From this revolution toward probing the unseeable, X-ray astronomy was born. The X ray is a very short wavelength, high energy radiation that has found many practical uses in medicine and technology. But natural X rays are alien to our everyday experience because our atmosphere totally absorbs them before they near the surface of earth.

The new X-ray astronomy frontier brought teams of scientists and engineers together for devising ways of sending X-ray detectors above the protective atmosphere. First they flew balloons carrying only Geiger counters. Then rockets probed higher, but only for minutes at a time. In 1963, the first rocket-borne experiments discovered a strong X-ray source in the constellation Scorpius. So intense were X rays from this location (named Scorpius X-1), that they were suspected of producing measurable changes in the earth's ionosphere.

Seven years later, X-ray detectors aboard

the satellite Uhuru opened floodgates of new data. From Uhuru's accurate sensors, scientists found more than 150 X-ray sources scattered throughout the sky. Since many of them were clustered along the Milky Way, astronomers concluded that at least eighty X-ray locations were within our galaxy. The remaining sources seem to be spread randomly over the rest of the sky, probably far from our Milky Way galaxy.

These suspected extragalactic points were studied closely and photos of the regions were examined. Twenty of them have been identified as clusters of galaxies — groupings in which 100, sometimes 1,000, galaxies swarm in an enormous cloud. Eight other sources overlap positions of individual galaxies, often unusually active ones. Some X-ray associated galaxies are exploding, and two or three X-ray spots are known to be quasars. At least forty other extragalactic sources remain unidentified; no connection has yet been found between the energy point and an observable object.

What about those galactic sources lying along the Milky Way? The location called Cygnus X-1 is almost certainly a black hole, and Hercules X-1 marks a double star system displaying periodic fluctuations in X ray and light emissions. At least twenty-four X-ray emitting binary stars are listed in the catalog of Uhuru sources — not all of them so dramatic in character as the above two. Five sources identify supernova remnants — the graves of dead stars. Four globular star clusters also emit X rays. That leaves a lot of X-ray sources unaccounted for. Perhaps the remaining forty-five hide behind obscuring clouds of interstellar dust.

Why do these various objects release X rays? In the case of binary star systems, the black hole or neutron star component sucks in gas from its companion star. As the matter spirals into the gravitational whirlpool, it gets tightly packed and heats to a plasma of millions of degrees (hot enough for emitting X rays). For matter plunging into a black hole, X rays are the last signal that can escape before the plasma vanishes.

The Crab nebula, the ghostly remains of a star that was seen to explode in 1054 A.D., contains two X-ray sources. One is nearly the size of the nebula itself, and at the center lies a pulsating X-ray object or pulsar. High energy particles spiraling around the nebula's magnetic fields are energetic enough to release X rays along with light and radio. X rays appear to be coming mainly from the large visible filament of matter near the pulsar (see chapter 17). The pulsar itself, that supercondensed core of the star that died in 1054, spins 30 times a second at the Crab's center. Some astronomers think that its dizzying rotation may provide energy for spewing off X rays (plus light and radio), but the processes are not yet understood in detail.

Apart from small X-ray sources that are probably mostly binary star systems, there is a concentration of X-ray activity toward the galactic center (in Sagittarius). This special location is clearly not starlike. Measuring tens of light-years across and centered at the Milky Way's core, this source suggests that at our galaxy's heart, something very energetic is occurring –- possibly an explosion, small by comparison with the distant X-ray sources and quasars.

In contrast to the mostly tiny X-ray

beacons within the Milky Way, X-ray regions associated with clusters of galaxies (regions of space containing dozens to thousands of galaxies clustered together) are as large as the clusters themselves. This means that the radiation does not come from simply one galaxy, but comes from something filling the volume of space between galaxies. Until that discovery, no one was sure if matter existed between galaxies; intergalactic space was thought to be empty. Today we know that is not so. The formation of stars from interstellar matter cannot possibly use it all up; material must be left over after star birth. So why the surprise at the X-ray indications of intergalactic matter? Mainly because all past searches employing radio and optical techniques have been unsuccessful. So far, intergalactic "starstuff" only becomes visible when astronomers extend their senses into the X-ray region.

The galaxy clusters in the constellations Coma and Perseus both emit X rays. The Perseus cluster source centers on NGC 1275, a bright, elliptical galaxy located almost at the cluster's core. Perhaps by some unknown process, NGC 1275 heats surrounding matter into radiating X rays. Although the intergalactic gas must approach nearly 100 million degrees Kelvin to produce X rays, very little matter is required — about one atom in every 200 cubic centimeters of space. However, the space between the galaxies in the Perseus cluster is so enormous that the total amount of gas is about four times the mass of the galaxies themselves.

Radio observations of NGC 1265 offer other proof for intergalactic matter. Data reveal a bright radio source at the galaxy's visible position and a radio tail streaming behind it. Like the wake from a drifting sailboat, the tail may be a disturbance left behind as NGC 1265 plows through the intergalactic medium. The galaxy itself seems to be throwing matter outward as a result of some interior explosion.

The discovery of matter between galaxies means that the universe is more massive than was originally thought. By measuring the total amount of mass in space, astronomers hope to learn if the universe is finite or infinite. Today we know that all galaxy clusters are rushing away from each other like material flung outward from an explosion. Will this expansion stop someday? A relatively massive universe would have sufficient gravity to brake the runaway galaxies, making it finite. But a lightweight universe would continue expanding forever, filling an infinite volume of space. Since the amount of visible matter in space falls ten to 100 times short of making the universe finite, astronomers have been searching for invisible or missing mass. The X-ray glowing stuff in the Perseus and Coma clusters suggests that the universe is twice as massive as was thought ten years ago. However, this remains far from enough mass to keep a lid on things.

Finally, on the largest scale imaginable, the entire universe emits X rays. This remarkable discovery was made from Uhuru when its telescopes pointed at the moon! The satellite's detectors effectively count the number of X rays hitting them every second. When they pointed at an X-ray star or a cluster of galaxies, the count rate increased — the number of counts being a measure of an X-ray source's strength. When the X-ray

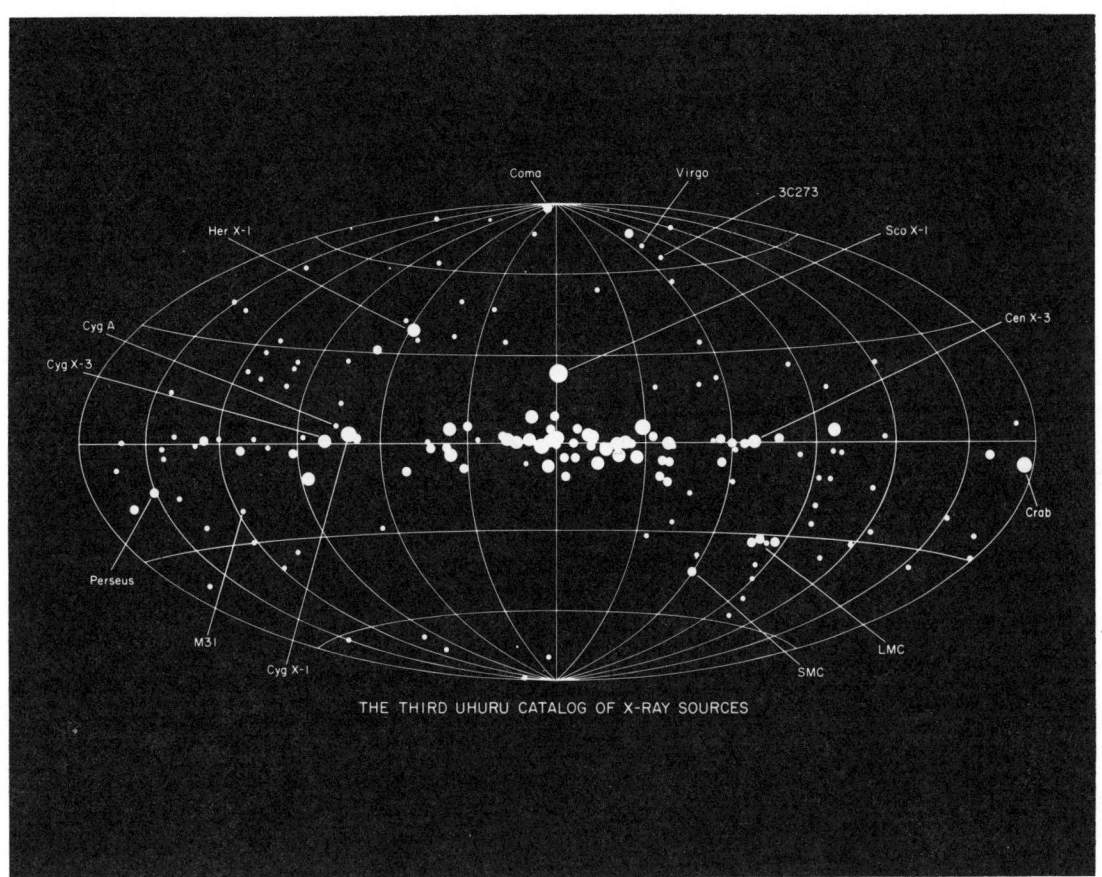

The location of X-ray sources based on observations by the Uhuru satellite are plotted on a galactic coordinates map with the plane of the Milky Way located horizontally across the center of the map. Much activity clusters along the Milky Way, indicating that many X-ray objects are within our galaxy. The brightest object in X-ray skies lies in Scorpius.

detectors pointed at the moon, the counts quickly dropped — implying that the moon shields X rays coming from distant parts of space. This always happened no matter where the moon was with respect to background stars and galaxies, meaning that the universe all around us radiates X rays.

Astronomers currently believe that these background X rays are ghostly signals still reaching us from a time shortly after the universe was created in the so-called big bang. In the mid-1960's, the "glowing embers" of that explosion were first detected. Although most of the radiation lies in the short radio-wave region of the spectrum, perhaps X rays account for a small percentage.

During that ancient event, the big bang, ultrahigh frequency gamma rays comprised most of the fireball's sizzling energy. So why is it diluted into microwaves and X rays now? Because since then, primordial radiation has been expanding into space along with the rest of the universe. As the light from external galaxies shifts to lower, redder wavelengths, the faster those galaxies recede from us (chapter 28); fireball energy should similarly drop in frequency as distance increases. Since this radiation was created long before galaxies, it must be reaching us from extreme distances of at least 12 to 15 billion light-years — implying great shifts in energy level.

Alternatively, the X-ray background may be produced by countless, very distant sources — all smoothed together by the relatively crude X-ray telescopes on board the satellite. This is unlikely, but only much more refined X-ray experiments will prove this beyond any doubt.

Through all evolution man has been almost totally blind, his vision restricted to one thin slice of the electromagnetic spectrum. Now spaceborne mechanical eyes are giving us almost godlike perspectives of objects and events invisible less than ten years ago.

SECTION III
INTERSTELLAR SPACE

Most of the volume of our Milky Way galaxy is an apparently empty void between the stars. However, it is not truly empty. Space between the stars contains gas, dust, magnetic fields, and nebulae of all shapes and sizes. Clouds of complex molecules are found around young stars and near other stars that are being born out of the interstellar material. When stars die they return material to interstellar space, continuing a recycling process. Hot stars, recently born from clouds of interstellar matter, heat up the surrounding gases so that they shine with their own light, producing many of the beautiful nebulae we see in photographs of the sky. The study of interstellar matter is of fundamental importance to astronomy; it is in interstellar matter that processes leading to star and planetary formation take place.

CHAPTER 8
BETWEEN THE STARS

"Surely these are holes in the heavens!" At least that's what William Herschel thought 200 years ago when he trained his telescope on dark areas in the Milky Way that appeared to be devoid of stars.

In 1909, E. E. Barnard, the most well-known observer of these "holes," finally accepted the fact that they were not holes at all but obscuring clouds of matter that blocked the light from more distant stars. Soon other astronomers accepted the notion that clouds of dust exist in interstellar space. But what are these clouds and where do they come from?

Photographs of nebulae almost always reveal the presence of dust clouds, and even the naked eye can detect dark lanes running through the Milky Way in Cygnus. These dark lanes are enormous clouds of interstellar dust. The presence of dust in interstellar space is a hindrance to astronomy because it prevents us from seeing very far into space along most of the Milky Way. In some directions we can only see 1,000 light-years away, but in others — like into the spiral arm in the constellation Carina in the southern hemisphere — it is believed that we can see for some 30,000 light-years at least. For some reason Carina is very clear of dust. (Dust clouds contain a lot of gas, but not all interstellar gas clouds contain dust.) Some clouds, like those in Cygnus, are very large and diffuse; others are small and dense, allowing no starlight at all to pass through them. These smaller clouds are proving to be particularly interesting in the light of recent research. They can range from small fractions of a light-year across to many light-years in diameter, and can absorb much of the light from a star that appears to be behind them.

It is possible that many of the smaller

INTERSTELLAR SPACE

The emission nebula M 16 in the constellation of Serpens. The hot incandescent gas is surrounded by dust clouds. Several very dense and totally dark dust clouds hang suspended in the foreground.

BETWEEN THE STARS 51

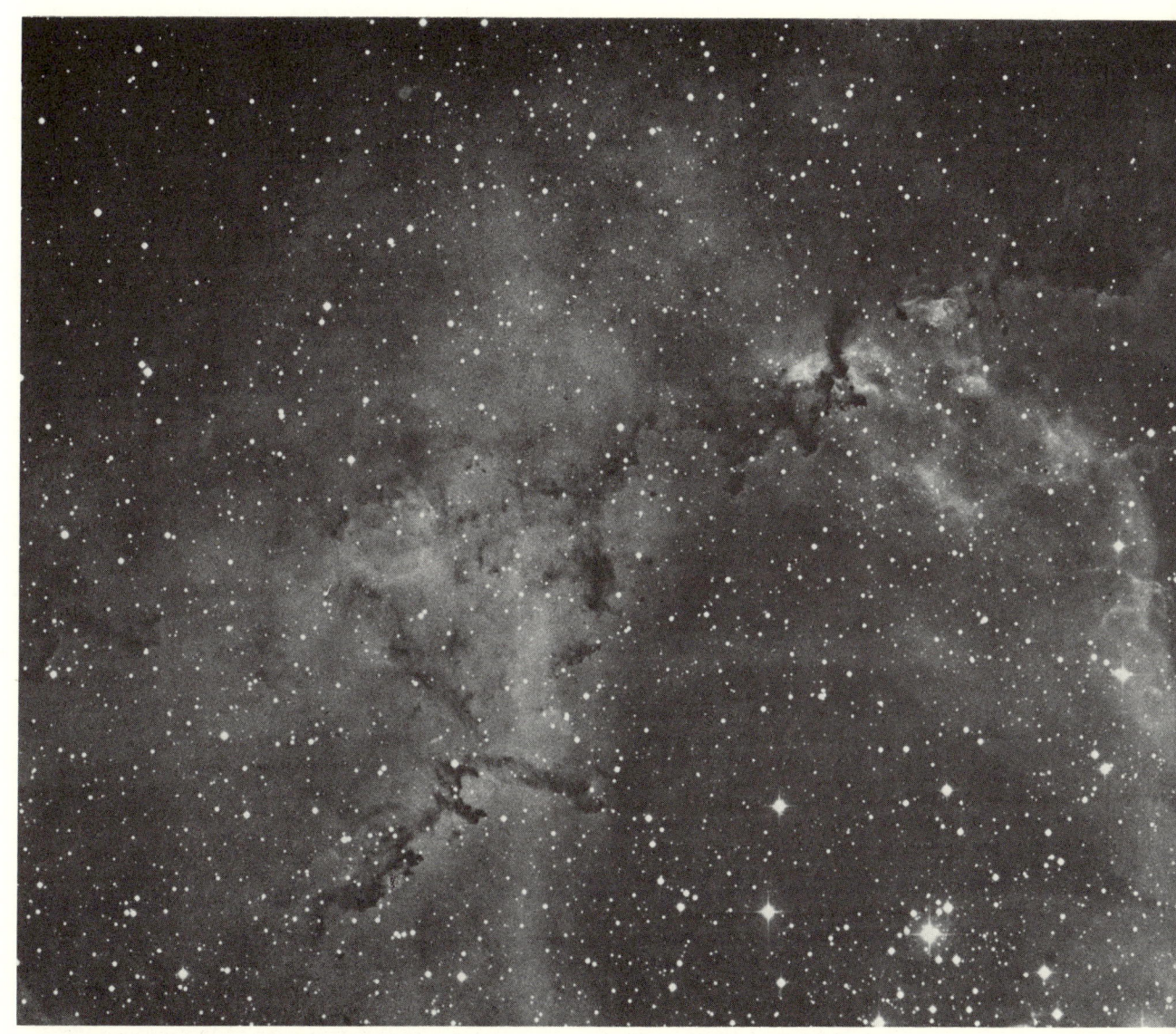

An enlarged view of the outer sections of the emission nebula known as the Rosette nebula showing the very dark dust clouds, known as globules, that totally cut off light from behind them. Many interstellar molecules are found in such clouds and they may often become birthplaces for new stars.

clouds, called globules, are in fact very dense clouds of matter collapsing to form stars. From optical observations of absorption of starlight we know that sodium and calcium atoms exist in the dust. By using radio astronomical techniques, many different atoms and molecules have been found to reside inside these clouds. Because molecules are found in the clouds, it means that the clouds are actually opaque and much of the radiation from space, which could destroy the molecules, is being effectively kept out. This, in turn, means that the density of dust particles is therefore large enough to absorb all the light and ultraviolet radiation. It's something like a rifle being fired into a dense forest. Before the bullet travels very far it will lodge in a tree. So it is with radiation entering one of these globules.

By astronomical standards these black clouds are dense — up to a million atoms or molecules per cubic centimeter. This compares with only one to ten per cubic centimeter in average interstellar clouds. Bear in mind that the best vacuums created in earth's laboratories cannot even approach the densities of these densest dust clouds. This makes experiments attempting to precisely duplicate conditions in dust clouds impossible to perform. It also means that astronomers have to do some very sophisticated, and therefore often incorrect, guessing as to the nature of the dust particles themselves.

The dust particles are known to be very small, probably about 0.00001 centimeters in diameter and about as reflective as snow. This is inferred from the way they actually absorb light at different wavelengths. They are believed to be elongated so they can produce polarization of starlight. (For a discussion of polarization, see chapter 9.) Some astronomers think the dust particles are graphite covered with a layer of ice; some believe silicates are the basic constituent; some have suggested that small diamonds would do the trick — the trick being to explain the observed effects of the dust on starlight. Interestingly enough, although we might not be certain of the precise nature of the dust particles, we can nevertheless go ahead and learn other useful things about the clouds. For example, where do they come from?

It is generally believed that the dust is produced in the atmosphere of giant stars. To examine this we will step back to an earlier stage of the evolution of the universe. The first stars that were born grew out of gas clouds with no dust in them. (Let's forget for the moment the theoretical problems associated with the cooling and contraction of such clouds in the absence of dust.) During certain stages in their evolution these stars became bloated and unstable, forming complicated molecules in their atmospheres and then puffing them off into space. These stars are effectively generators of the dust particles. We know this to be true since we can observe strong stellar winds blowing away from many giant stars and we can recognize the presence of the material in these winds by its absorption of the star's normal light. Therefore, as the galaxy evolves, the interstellar medium gathers an increasing amount of dust from these stars. There seem to be enough of the right kind of giant,

BETWEEN THE STARS 53

A very dense cloud of dust whose shape is clearly defined where there is total darkness on the photo. No light from more distant stars can penetrate this dust cloud, which is a site for many interstellar molecules discussed in this chapter. This dust cloud is known as Barnard 335.

cool stars to have produced the clouds we see, so this scenario of dust formation is probably correct.

The newly formed dust clouds are ideal places to build molecules, many of which are needed to form life on planets. So, as the universe evolves, molecules of increasing complexity are capable of being formed and, more importantly, are capable of being preserved within the dust clouds. We also know that many of the globules are presently regions of star formation; today it is a source of endless discussion as to how many of the molecules found there (such as water, formaldehyde, ammonia, carbon monoxide, and hydrogen cyanide) might survive the star birth process and remain in the surrounding nebula that will form planets. This is a fascinating question because the molecules are the ones that were needed for biological structures to evolve on primitive earth.

The globules therefore are the state of matter just before star formation. Because they are dark and cold (often only a few degrees above absolute zero), the pressure of hotter surrounding matter on their outer surfaces tends to cause contraction, thus increasing the density. Eventually the density reaches the point where individual stars, and presumably planetary systems, emerge. Many stars and planets could be formed because the amount of matter in a typical cloud is considerable despite its low density. Often globules are several light-years across and contain enough matter to give birth to many thousands of stars like the sun.

Let us look a little more closely at what occurs inside dense dust clouds. If we accept the fact that there are small dust particles in the clouds, then we can use these to construct additional molecules in the following manner. Take two atoms and call them X and Y. We want to make the molecule XY. One way we can do it is to simply wait until X and Y collide by chance in the cloud. Of course the denser the cloud, the more likely this is to happen. On the other hand, X can bump into a dust grain that is much larger than X or Y, and stick to it. Then later, Y does the same and they may slowly migrate across the surface and form XY. But astronomers would not be able to observe the true nature of XY, or even its presence, if it stayed stuck to the grain. Somehow, before we can detect it, it has to get loose again so that it can float off and join up with all the other XY's. To get it unstuck we have to give it a little energy. But then we find that not only do we free XY, we also break XY up into X and Y separately again! This problem is a serious one that prevents an easy explanation for the existence of molecules in dust clouds. Even when the average XY is found floating about in the cloud, radiation leaking into the cloud from outer space is sufficient to destroy a typical XY every thirty years. Yet the clouds are millions of years old, and still they are full of XY and molecule building occurs all the time to balance the destruction.

The processes required to build molecules in clouds are only just beginning to be understood. They involve starting off with a carbon atom, shining light on it, ionizing it, making it collide with molecular hydrogen, shining more light on it, and letting the resultant combinations collide with electrons, protons, other atoms, and so forth, to allow the building of many complex molecules to counter their destruction. This theoretical examination of molecule building is now part of the new field of astro-chemistry that has emerged as a consequence of the recent detection of molecular clouds in the "holes in the heavens." It is still in its infancy, but astro-chemistry holds the promise of tying together stars, life, and evolution — an awesome arena for the minds of humans.

CHAPTER 9
MAGNETIC FIELDS BEYOND EARTH

We all know what a magnetic field is — that strange invisible force that causes a compass needle to point north and attracts two lumps of iron to one another even though they aren't touching.

Earth has a magnetic field. Without it compasses wouldn't be of much use. Earth's field has changed its direction during the history of our planet and these changes are permanently recorded in the magnetization of rocks deep below the surface. The strength of magnetic fields is measured in units of Gauss, named after the physicist who significantly increased our understanding of this strange, invisible force. Earth's magnetic field has a strength of about three-tenths of a Gauss at the surface of our planet.

The sun, too, has a magnetic field. So do Jupiter, Saturn, and Mercury, but Mars does not. The moon has an extremely weak field. On the sun, the field strength is about one Gauss, but around sunspots magnetic fields have strengths of hundreds of Gauss. These strong fields determine the motions of hot gases in the solar atmosphere, and the gases contain electrons that are affected by magnetic fields; moving electrons can generate their own magnetic fields. So it is in the sun.

The beautiful loops and curves in solar prominences are usually produced by the structures in the magnetic fields above sunspots. The hot clouds of gases, mostly electrons, are forced to stream along the so-called lines of the magnetic field. They rush upward, and when their energy is expended they fall back down to the surface of the sun, again following the magnetic field lines. Astronomers can learn a lot about the solar magnetic fields in and around sunspots, and daily maps of the solar magnetic fields are produced at several observatories.

But do magnetic fields exist in space beyond earth and the sun? Surprisingly, they are everywhere. They thread through space between the stars; they permeate dust clouds, forcing some of them to take on an elongated appearance; and they build up in strength as interstellar clouds get denser during their evolution into stars. However, at some point this increase in magnetic field strength should prevent the cloud from condensing any further, because magnetic forces tend to counteract the effects of gravity. Yet stars have formed, so for some reason not yet explained, magnetic fields must escape the protostellar cloud before the star can be born. Of course there are so-called magnetic stars which have fields of thousands of Gauss, but they too must have lost some of the original fields that would have built up during the early phases of contraction in the gas clouds.

How do we know that magnetic fields exist out there in interstellar space? Obviously we haven't sent a compass into regions of space many light-years from earth, so how can an astronomer become aware of the existence of something so insubstantial as a distant magnetic field? The answer lies in being able to detect the effects that invisible magnetic fields have on other phenomena or radiations.

As is true with so many astronomical discoveries, detection of the first interstellar magnetic fields was an accident. The discovery arose out of an experiment searching for polarization of light from certain stars that are members of binary systems. First, a word about polarization, a concept that is often difficult to envisage.

We are all familiar with the action of polarized sunglasses that cut the glare on a sunny day. Rotate such a pair of glasses and varying amounts of light appear to be transmitted through to your eye. Set up two pairs of polarized glasses in front of one another, rotate one by 90 degrees, and no light will be transmitted at all. This is because the lenses are actually polarizers. The first pair only allows light vibrating in a certain direction to pass. The next pair would do the same except that if it is rotated by 90 degrees it will only allow light polarized in the new direction to pass. But the first filter already cut that light out, so nothing is left. The point of using polarized sun glasses is that light from the sun, originally unpolarized, becomes polarized on reflection from certain surfaces on the the ground. The polaroid filters are therefore lined up so they cut out the reflected light (glare).

Now, if you constructed a high-quality polarizing filter, placed it in front of your eyes, looked at the stars, and then rotated the filter, you would discover that the brightness of the light from many stars would vary with the angle of the filter. This experiment is not practical since the eye is not sufficiently sensitive to the small amounts of polarization present in starlight, but such filters mounted on telescopes do allow astronomers to measure the polarization of starlight.

The first experiments to search for polarization of starlight were motivated by a desire to find out if processes in the atmosphere of certain stars produced polarization at the rims of the stars. This polarization

58 INTERSTELLAR SPACE

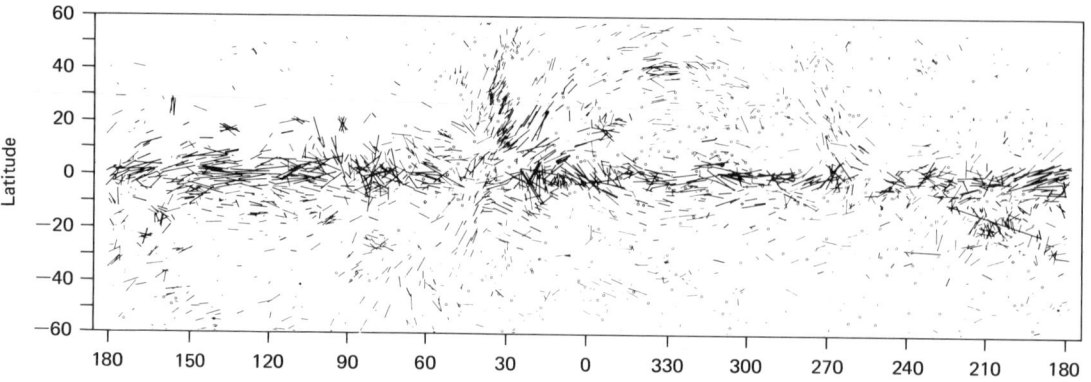

A map that indicates the direction of the magnetic fields that thread their way between the stars in our Milky Way. Each line represents the direction of the polarization of a star and the length of the lines indicate the amount of polarization. The patterns reveal the nature of the magnetic fields in space much as iron filings reveal a pattern of a magnetic field in simple laboratory experiments.

was supposed to follow the curvature of the circle defining the star's surface. It was thought that eclipsing binary stars (pairs of stars orbiting one another with each, in turn, coming in front of the other as seen from earth) would be the best candidates for finding the effect; instead, it appeared that single stars showed polarization, with different stars showing different amounts of polarization. The experimenters soon found that the amount of polarization correlated with how much dust lay between the star and the earth. Clearly, interstellar dust was affecting the light being received.

Today we realize that this is not unreasonable since we know that sunlight, reflected off the ground, becomes polarized. Similarly, starlight being reflected off interstellar dust particles can become polarized. However, in this case we don't see the reflected starlight, but only that part of the light that makes it through to us (the transmitted light); it, too, is polarized.

This explanation for the cause of the polarization of starlight led to a better idea of the make-up of the dust in space. For polarization to occur, we need elongated dust grains that are preferentially lined up by some mechanism. For this, a magnetic field is required, although some suggest that "winds" in space would line up such particles as well. If the magnetic fields do line up the dust particles and we measure the polarization of many stars, we should be able to make a map of the magnetic field in interstellar space. This has been done. A map of the polarization directions for thousands of stars is shown above. Each little line represents the direction of polarization and hence the magnetic field orientation in the dust clouds in the direction of the stars being measured.

What about the strength of the fields required to align dust particles? No one is certain, and it depends on the true nature of the dust grains; but fields around one hundred-thousandth of a Gauss should be enough.

Another way to measure magnetic fields in space involves examining the effect that a magnetic field has on radio signals from clouds of hydrogen gas. The experiments are difficult and so far only five measurements have been successfully made. They show fields of only several millionths of a Gauss in some clouds. That is a very weak field and exists in clouds many thousands of light-years from us; yet we can measure it.

Another direct indication of the presence of magnetic fields in space comes from observations of radio signals from the Milky Way. Interstellar space is full of cosmic rays — high energy electrons traveling nearly at the speed of light. When these cosmic rays find magnetic fields, they are forced into giant spirals around the magnetic lines of force. In doing so they lose energy by radiating it away as radio signals. This generates the radio static that Karl Jansky, back in the 1930's, discovered coming from the Milky Way.

Radio astronomers map the radio emission that comes from all along the Milky Way band and find that the radio signals are polarized. They measure the polarization of radio waves by rotating their antennas while pointing at the source. You can do such an experiment with your television or FM radio antenna. If you rotate the antenna from the horizontal to the vertical position, while still pointing at the distant transmitter, you will find the received signal weakens considerably. This is because the transmitter is sending out a horizontally polarized signal and your antenna is polarized along the direction of its maximum length. You have to place it horizontally to pick up most of the signal.

Supernova remnants also contain strong magnetic fields. As the shell of matter from the explosion expands away from the exploded star, magnetic fields in space around the original star build up so that the shell actually acts as an amplifier of the surrounding magnetic fields. In addition, the explosion generates a large number of high energy particles — we would call them cosmic rays — and these rush outward and meet the surrounding regions of compressed magnetic fields. The result is that radio signals and light are emitted by the cosmic-ray particles as they spiral about the field lines. Hence, we can see such objects as the Crab nebula and also pick up radio waves from it.

Light from the Crab nebula is also polarized and photographs taken with the polarizing filters at different angles reveal this quite clearly. In the case of this emission, the direction of the polarization of the light is at right angles to the field lines. Examination of the photograph of the Crab nebula on page 60 shows elongated structures at right angles to the angle of the filter, because it is the emission from cosmic rays spiraling about those particular magnetic field lines which we see. The magnetic fields in the filaments of the Crab nebula are about one thousandth of a Gauss in strength.

Radio signals from distant radio galaxies and quasars are also polarized and magnetic fields in those objects are at most a few ten-thousandths of a Gauss. But none of these objects is a record holder as far as strong

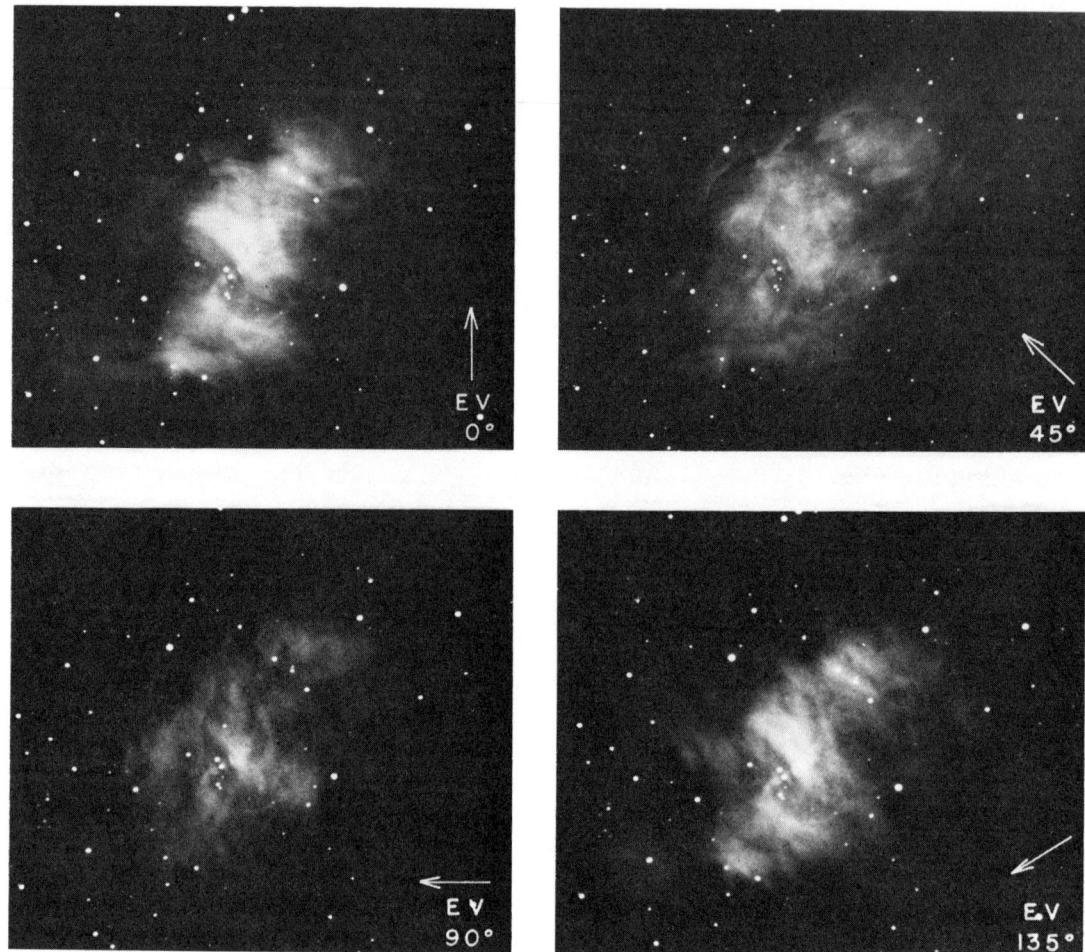

The Crab nebula, the remains of the star that was seen exploding in 1054 A.D., as photographed in polarized light. The direction of polarization of the filter used is shown by the arrow; it is obvious that at different angles, different parts of the nebula are seen based on the differences in structure in the photos. This supports the theory explaining the emission of light from the nebula, which requires the presence of magnetic fields and high energy particles. The magnetic fields are revealed in the striations they produce as a result of their controlling influence on the way matter moves within the nebula.

magnetic fields in astronomy are concerned. That distinction is held by pulsars.

To explain the incredibly regular pulses of light, radio, or X-ray emission from pulsars, astronomers have had to invoke a complicated picture of what a pulsar is. Basically it is an incredibly dense 15 kilometer wide sphere called a neutron star. As it rotates, usually several times per second, it carries an intense magnetic field (a trillion times stronger than that of the earth) around with it.

In addition we have to have something asymmetrical occurring so that we only get one pulse per rotation. The way to do this is to put the magnetic field in the spinning star with its poles offset from the axis of rotation, just the way earth's magnetic north pole is not at the true north pole. If this situation exists in the pulsar, then we should regularly see each magnetic pole of the pulsar facing us once per revolution. This has allowed astronomers to estimate the magnetic field strength of pulsars. The answer comes out to a staggering one trillion Gauss. This means that astronomers are observing cosmical magnetic fields ranging from one millionth to one trillion Gauss. There are probably few other single physical parameters being studied today that show such an enormous range of values.

CHAPTER 10
JEWELS OF THE UNIVERSE

Everywhere swirling clouds of hydrogen gas and dust drift and stream through interstellar space. Sometimes stars are born in these clouds. When this happens, they heat up the remaining surrounding gas and produce a glowing cloud visible to us on earth. These clouds are known as emission nebulae, by virtue of the fact that they emit their own radiation.

Let's abandon our vantage point on the surface of our planet and travel out to one of these incandescent gas clouds to see what they are like up close. Forget for a moment, that nebulae such as the Orion (see page 70), Trifid (see page 63), Lagoon (see page 66) and Hourglass are so far away that they appear small in a telescope. Instead, let your imagination wander out toward them and see them as enormous, extremely hot clouds — breathtaking nebulae in which delicate, dynamic atomic processes are occurring.

Millions of years ago, a dense cloud of cold hydrogen gas containing dust particles in profusion began to collapse slowly under the influence of its own gravity. Small pockets of dense material within the larger cloud began to collapse at a faster rate, heating up due to the pressure of gravitational collapse. These small pockets ultimately turned into protostars that grew very hot (1,000 degrees Kelvin) and dense. Today, the remaining dense cloud of dust surrounding these "baby" stars does not allow their visible light to escape, but the infrared (heat) radiation can penetrate the dust; this is why we can detect protostars from earth. One or more protostars probably exist behind the Orion nebula (chapter 11), and many other infrared sources have been

JEWELS OF THE UNIVERSE   63

The Trifid nebula, located in the constellation of Sagittarius, in which young hot stars were recently born and now heat the surrounding gas to incandescence.

found by astronomers surveying the sky with infrared telescopes located atop high mountains. The earth's atmosphere absorbs infrared radiation, so astronomers try to get above as much of it as possible (see chapter 5).

As our protostar grows even hotter and smaller, its radiation destroys or drives away the surrounding dust so that 100,000 years later, its light can escape out into space. A star we can see is thus born. Such a new star is often very hot and very large. Most importantly, it radiates a tremendous amount of short wavelength ultraviolet (UV) radiation.

The UV radiation from the star encounters cold hydrogen atoms in the gas cloud that surrounds it. When this happens, the radiation tears away electrons from the original hydrogen atoms and we then say that the hydrogen is ionized.

When the electrons are knocked free, they rush through the gas cloud at many thousands of kilometers per second. These velocities are equivalent to temperatures of many thousands of degrees. Thus UV radiation from the hot, young star is converted to energy in the gas cloud surrounding it, which effectively heats up the gas cloud.

As these free electrons hurtle through the cloud, they occasionally meet protons on similar wild paths. Such encounters cause the electrons to lose, or radiate away, some of their energy in the form of optical and radio electromagnetic waves. This is due to the fact that the electron is a charged particle that, when accelerated (changes its speed or its direction of travel), will generate radiation at some particular wavelength. The fact that signals are generated over a wide range of wavelengths, from visible light to radio waves (section II), is due to the existence of an enormous number of electrons being involved in this process simultaneously. Encounters between the electrons can take on a whole range of values, from direct collisions to very distant, weak interactions. The density of the gas cloud determines the frequency of such electron encounters, and the frequency and nature of the encounters between electrons and protons, as well as the velocity of the electrons, determine the range of wavelengths radiated and the intensity of the signals emitted.

Let's look closely at the region around one of the hot stars that has just been born in the larger cloud. Imagine the star to be just in front of you, and imagine that the gas cloud surrounding it is perfectly uniform in density. UV radiation emitted from the young star first encounters and then ionizes the hydrogen atoms in the star's immediate vicinity. Radiation that follows is free to move unhindered through the already ionized material. When this radiation finally encounters and ionizes atoms, it is also absorbed and not able to ionize any further material. This process continues outward through the cloud to greater and greater distances from the star.

However, the process does not continue indefinitely — a counterbalancing effect also takes place. Electrons and protons are recombined to form hydrogen atoms that may, once again, be ionized by more UV radiation. A point of balance is reached when the ionizations and recombinations are equal in number. It is at this point that the sphere ionized matter around the star be-

The emission nebula IC 2944, a region of recent star formation where dust, gas, and young stars coexist. A number of very tiny dust clouds, totally black spots against the background of the nebula, can be seen. These are examples of globules, thought to be sites for future star formation.

The Lagoon nebula, an emission nebula laced by tenuous dust clouds that cut off the light from the nebula. Some very tiny and very dark dust clouds, or globules, can be seen near the edges of the nebula.

comes no larger. Recombinations inside this sphere produce enough additional atoms to absorb all the UV radiation emitted by the star in a given volume of space.

The Danish astronomer Bengt Stromgren was the first to work out a detailed theory of this process, and the ionized sphere of matter surrounding a star located in a uniform gas cloud is referred to as the Stromgren sphere. The radius of the sphere is dependent on both how many atoms are capable of absorbing the UV radiation and the temperature or UV output of the star. For very hot stars, the radius of a typical cloud can be tens of light-years, but for a star like the sun it may be extremely small (fractions of a light-year).

Nature does not usually produce single stars in uniform clouds of gas. Clouds will have considerable density variations within them, and usually many stars are born simultaneously in such a cloud. For this reason, emission nebulae photographed from earth show all sorts of beautiful shapes and sizes — shapes and sizes that depend on the structure that exists within the cloud in which the stars are located. The emission from the hot gas is generally blue in color, and the starlight reflected from the dust within and around the nebula tends to be red. The resulting color patterns in a nebula can be quite staggering in their beauty.

How can we be certain that recombination of the electrons and protons is really occurring? We can observe these recombinations directly because of the radio signals that are emitted during recombination.

When an electron drops into an orbit closer to the proton, we find that the atom's total energy is less. In the case of hydrogen atoms, this energy difference can be radiated away from the atom as a radio signal. When the electrons drop from the very highest orbit in the hydrogen atom to the next lowest, energy emissions are in radio frequencies. Each change in level involves radiation at one frequency only, thus producing what is known as a spectral line. Changes in orbit among the lowest levels are visible at optical wavelengths, producing spectral lines such as the well-known hydrogen alpha line. Recombining hydrogen atoms produce about 200 spectral lines between radio wavelengths of 1 and 75 centimeters.

Study of the recombination spectral line shapes and intensities, as compared with the emission that covers all optical and radio wavelengths, enables the astronomer to derive information about a gas cloud's temperatures and densities. On the other hand, study of the frequencies of the observed lines as compared with the expected values allows one to determine the atom's motions within the nebula. A shift in frequency is produced by motion in the cloud, a phenomenon known as the Doppler effect.

When we look at a photograph of an emission nebula, such as the Lagoon nebula in Sagittarius, we often see dark blobs projected on a brighter background. Some of these dark blobs show wisps of bright material radiating toward the ionizing stars. This is matter "boiling" off the surface of the dark blob and streaming back into the nebula. These dark blobs are very dense dust clouds into which the ionizing UV radiation cannot penetrate. However, the surface layers are very slowly being worn away and may stream out from the dense dust globule and stream toward the less dense region of space around the hot stars. In some cases, the dust cloud is very small and sufficiently dense to produce a shadowing effect on matter more distant from the ionizing stars. One then sees a long tongue of dark material projected against a bright background, with the tongue pointing directly away from the nearby ionizing star. Such objects are often referred to as elephant trunks because of their distinctive appearance.

All these dynamic processes occur in such jewels in the sky as the Orion, Lagoon, and Trifid nebulae. Next time you look at one of these objects with the naked eye or telescope, try to picture it up close. Imagine the delicate balance between ionization and recombination that determines its detailed appearance. Think of the numerous spectral lines radiating toward us that indicate the recombination, not only of the hydrogen atoms, but also of helium and ionized carbon. And try to imagine the enormous extent and great temperatures that exist in these regions of space, where stars are still being born.

CHAPTER 11
INSIDE AND BEHIND THE ORION NEBULA

Some winter night look at the faint middle star in the sword of Orion. You'll see that it appears somewhat fuzzy. If you direct a small telescope to this region you'll see a cloudlike patch with a few stars embedded in it. In a larger instrument the cloud's shape is more clearly defined and greenish wisps make it a fascinating sight on a dark moonless night. When Galileo drew this region of the sky in 1609, he either did not notice the nebula or his telescope was too crude to show it.

Early investigators of the Orion nebula could not have imagined its magnificent, swirling beauty — fully revealed only in long-exposure photographs. Indeed, during the 20th century, the Orion nebula has proven to be one of the most photographed, and most interesting, objects in astronomy. In recent years a great deal has been learned about conditions within and immediately surrounding this nebula.

The Orion nebula is one of the youngest objects in the sky. If cave dwellers some 25,000 years ago ever plotted the stars of Orion (the Hunter) they would have omitted the central star in the sword. The nebula is believed to have started shining about 23,000 years ago. It is a relative newcomer on the celestial scene.

If our sun was ten times closer to the nebula (140 light-years instead of 1,400), it would be the most spectacular object in the sky. It would cover an area equal in size to most of the constellation Orion and be as bright as the brightest stars we can see. Despite its vast distance, new techniques are adding pieces to the Orion nebula story. Even its age is now known. The new findings show that the nebula lies directly in front

The Great nebula in Orion, located about 1,200 light-years away in the sword of Orion, is shown here in a dramatic photograph produced with the 158-inch telescope at Kitt Peak. The dark dust clouds thread their way in front of the hot gases surrounding the Trapezium stars.

INSIDE AND BEHIND THE ORION NEBULA 71

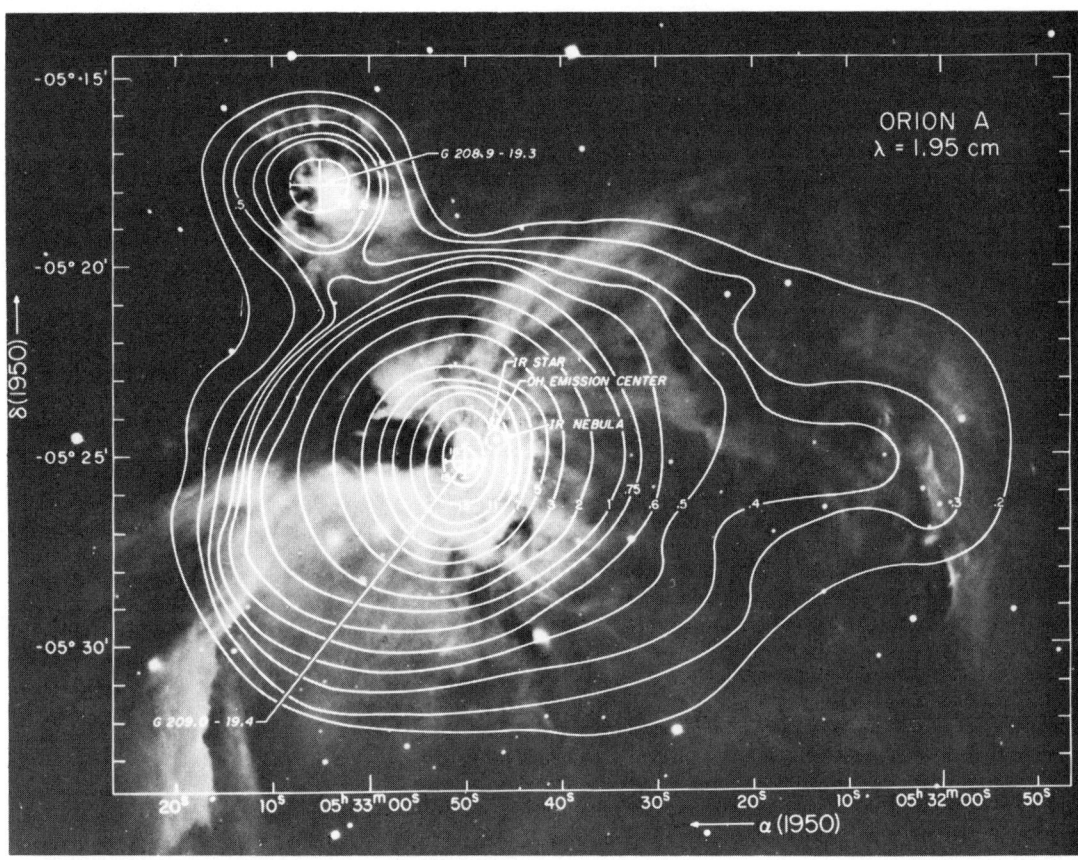

A radio map of the source associated with the Great nebula in Orion superimposed on a photo of this nebula. The location of infrared sources of emission as well as a source of OH spectral line emission are indicated.

of a very dense cloud of gas and dust. This dense cloud contains many complex molecules and is believed to be the scene of star formation.

The Orion nebula itself is a hot cloud of incandescent gas called an emission nebula because it emits its own light and radio signals due to the high temperature of the gas in the cloud. The nebula's temperature of 8,000 degrees Kelvin is due to ultraviolet radiation from a group of four hot massive stars called the Trapezium.

Photographs of the Orion nebula show that it consists of two patches — the regions of emission. Radio maps made at many different wavelengths also show this double nature that indicates that hot gases are distributed in these two complexes. Studies of the radio maps have confirmed that the process of emission giving rise to the radio signals is the so-called thermal process, due to near collisions between electrons in a hot gas.

The radio observations of the molecular radiation (see chapter 12) from the direction of the Orion nebula turned out to be the most exiciting recent investigation. The nebula we can see lies in front of a molecular cloud whose center is lined up just below and behind the Trapezium. The dimensions can even be figured out by the correct interpretation of the enormous funds of data now available for this region. The giant molecular cloud lies one light-year behind the Trapezium. The inner part of this cloud has a diameter of about 1.5 light-years. The cloud extends further than this, but its actual outer boundaries have not been accurately determined. The visible nebula is about 3 light-years deep in our line of sight, about 6 light-years across, and has a mass equal to 100 suns.

There is relatively little dust in the nebula itself, and what dust there is appears to have come from the nearby, but hidden, dusty cloud of molecules. There is so much dust in the molecular cloud that it is totally opaque to light.

Within the molecular cloud is a strong infrared object (see chapter 5), about 4 minutes of arc in diameter (about one light-year), called the KL nebula. Near its center is a tiny infrared star (called Becklin's star). Neither of these is visible optically and both were found in surveys of the sky made with infrared telescopes. It now seems pretty certain that these two objects indicate the presence of protostars in the molecular cloud. The infrared star has a temperature of about 600 degrees Kelvin, and its infrared radiation can leak out through the dense dust cloud that surrounds it but its light radiation cannot.

The problem up to now has been to explain the existence of this infrared star and the KL nebula. It was thought that possibly the star was such a strong infrared source and yet invisible to optical astronomers because it lies behind an enormous amount of dust. In this case, the light does not shine through, but the infrared (the heat radiation) does come through. Alternatively, it was thought to be a very cool, protostar (600 degrees Kelvin) surrounded by a cocoon of dust. The star might heat this dust and the infrared thus radiated is what the astronomers detect. Becklin's star could also be a more normal star, partly surrounded by dust. Indeed, several infrared stars have now been found immediately around Becklin's star and these seem to form a cluster.

The new suggestion is that these stars are located on the other side of the molecular cloud, which lies just behind the Orion nebula, and there is much dust on the edge of the molecular cloud partly surrounding those stars. They appear as infrared stars, then, because all their light is absorbed in the dust cloud. The KL nebula is that part of the dust cloud heated by the stars and reradiated in the infrared.

To get a better picture of what is happening, imagine you are an observer located way on the other side of the Orion region. You would be able to see the star cluster that is invisible to us and from your vantage point the molecular cloud would lie behind those stars, and then would come the emission nebula we call the Orion nebula. You would not be able to see the Orion nebula at all, and the Trapezium stars would probably appear as four infrared stars! Apparently the Trapezium is but one of at least two clusters of stars that lie next to one of the most interesting clouds of complex molecules we have found in the galaxy.

At least ten molecules (including formaldehyde, hydrogen cyanide, and carbon monoxide) have been found by radio astronomers studying the molecular cloud behind Orion. The size of the cloud seems to depend on the species of molecule being studied. This is because the existence of a molecule and its ability to radiate depend on the physical conditions in its vicinity. For example, some molecules occur naturally only inside a dense dust cloud where they are shielded from harmful radiation from outer space, but more stable ones can exist perfectly well in the exposed outer parts of clouds. Maps of the molecular cloud therefore look different depending on which molecule is being studied. Detailed studies on the way molecules are distributed in the Orion cloud give valuable clues about the processes of astro-chemistry.

The study of the motions in the Orion region can be inferred from the velocities of the Trapezium stars, and by the molecular spectral lines and the recombination lines of hydrogen and carbon atoms. Interpretation of these motions gives even more information about conditions in that part of space. The observations would at first indicate that the emission nebula, the one we see in photographs, is approaching us while the stars and the invisible cloud of molecules are not. This is because the ionizing radiation from the stars is virtually boiling off the front side of the molecular cloud. The ionized particles stream away from the molecular cloud into the less dense regions of the Orion nebula itself. Hence, they are streaming toward us and give the appearance that the nebula is moving toward us.

We also know that there is at least one relatively dense cloud of hydrogen gas, somewhat smaller in size than the visible nebula, which lies in front of the nebula. This cloud absorbs the radio signals from the nebula at the hydrogen wavelength of 21 cm. It has the distinction of containing the largest magnetic field directly measured in interstellar space (about one ten-thousandth the strength of the earth's field).

Next time you look at the sword of the magnificent winter constellation of Orion, remember that it contains one of the best studied regions of space and that star formation is even now taking place behind the great nebula.

CHAPTER 12
INTERSTELLAR MOLECULES

We have been flooded with news items about the discovery of this or that molecule between the stars during the last several years (molecules are groups of atoms that form stable combinations). Where has it all led and what have we now learned about the chemistry of the space between the stars? One striking discovery is that the chemistry in space is much like the chemistry on this planet. In particular, the extremely important processes that are part of what we call organic chemistry occur in space. Organic chemistry deals with the reactions of molecules containing carbon atoms, and the chemistry of carbon is the basis of life on this planet. Now we discover that events in the very alien conditions in clouds of gas and dust between the stars also favor this chemistry above all else. If organic processes are so common, we can comfortably speculate that these processes will occur on other planets; hence, we can start to speculate about life on other planets and on how it might be similar to life on earth. In other words, should life evolve elsewhere it is likely that it will be of a form that we can readily recognize.

In a random and tediously slow way, molecules of increasing complexity appear to be evolving in the depths of space. The search for their existence has become one of the most competitive areas of modern astronomical research, and over the last few years we have been flooded with discoveries of new interstellar molecules.

In the early 1940's, three interstellar molecules were identified. They were CH, CN (cyanogen), and CH^+. Here, C refers to the carbon atom, N is the nitrogen atom, and H is the hydrogen atom. When the single electron orbiting the hydrogen atom is stripped from its orbit, the atom is said to be ionized and the notation H^+ is used. The three molecules were discovered by the effect they have on the light from distant stars; they dimmed the light from these stars at very definite wavelengths (or colors) and only at these wavelengths.

No new interstellar molecules were discovered until 1963 when the combination of the oxygen (O) and hydrogen (H) atom, called simply OH, was discovered because it emitted a radio signal at four characteristic wavelengths around 18 cm. Radio astronomers studied this molecule in various dust clouds and gas clouds in our Milky Way galaxy for another five years, not suspecting that there might be other more complex molecules in them. Then, in 1968, the dam burst. Astronomers at the University of California made the startling discovery that there were clouds in space containing water (H_2O) and ammonia (NH_3). These two molecules emitted radio signals at around 1.35 centimeter radio wavelengths and large radio telescopes had been used to find these signals. However, the radio signal from clouds of water in space was found to be so strong that anyone could have found the signal at least fifteen years before, if only anyone had thought to look for it!

Now, some seven years after the discovery of the water and ammonia molecules, astronomers have discovered that at least forty different molecular species exist in clouds of gas and dust in space. Radio signals from at least three other species, as yet unidentified, have also been found.

How can we be certain that a particular radio signal comes from a particular molecule? There are two ways to find out. First, the signature of a particular species can be measured in the laboratory. Second, using a large computer, it is possible to calculate what the signals should be if the structure of the molecule is accurately known. Basically, a particular type of molecule will transmit or absorb a radio (or light, or infrared) signal if its state of motion changes. For example, this motion can be vibration within the molecule or end-over-end rotation of the molecule. Sometimes, because of limits in present scientific knowledge, it is nearly impossible to calculate what these motions are likely to be or to identify them in the laboratory. This, therefore, prevents the ready discovery of many more complex molecules in space. In particular, we would like to know if the precursors of life, the amino acids, exist in the hostile environment of space, but no one has successfully been able to predict at what wavelength even the simplest amino acids might radiate or absorb energy. As science and technology progress the likelihood that someone will succeed in finding the characteristic "signature" of an amino acid will increase, thus allowing it to be detected in space.

The incredible thing is that many of the molecules that are regarded as fundamental for building the precursors of life on this planet are out there in space in huge quantities. These are molecules such as hydrogen cyanide, water, ammonia, methyl alcohol, and cyanoacetylene. Few astronomers will speculate about the relevance of this. The question is that while these molecules exist in the clouds of gas and dust in space, do they really make any difference to what subsequently happens on newly formed planets? Since the theories about the evolution of life on a planet require the

76 INTERSTELLAR SPACE

Two emission nebulae of interest to radio astronomers. The large diffuse object in the center of this print is the nebula IC 1805 and the small, bright object at the bottom right is called IC 1795. Known as W4 and W3 by radio astronomers, these nebulae emit strong radio signals and are sites for several clouds containing many OH molecules.

presence of all these molecules, their initial existence in the clouds from which the stars and planets form might possibly speed up the evolution of life on planets, assuming the molecules survive the process of planetary formation.

When a star condenses out of a cloud of gas and dust, which certainly must have contained most, if not all, of the molecules so far observed, the star heats up and starts to shine and its radiation probably destroys the molecules, breaking them up into their constituent atoms again. This is based on a simple picture of how stars, and ultimately planets, form. If, however, planets form out of the dust cloud that is left around the star, and if this dust is sufficiently thick, as some believe, then the dust will shield these delicate molecules from the destructive starlight; possibly many will survive through the processes of planetary formation. While this is not yet known, it is an interesting question, since we can then picture planets that are ripe for the start of life more or less from the time they are formed. The chances for the evolution of the type of life that is based on carbon chemistry throughout the galaxy and the universe are then very high indeed.

The other interesting discovery is that the most complex molecules have been found mainly in a few very unique clouds in the Galaxy. One of these lies behind the Orion nebula (see chapter 11) and the other lies in the direction of the center of our Milky Way (see chapter 21). This latter cloud is hidden from our view because of the enormous amounts of dust between the center of the Milky Way and us. At present, whenever an astronomer wants to search for a new molecule in space, the first places searched are in these two directions. It is likely that there are many other equally interesting clouds in the Galaxy, but they have not yet been found. It is also clear that within these clouds stars are probably being born even now. This is based on the detailed studies of these clouds using infrared wavelengths that allow one to see the very young protostars (see chapter 5). It seems a pity that this rich organic soup should all be destroyed when the star starts to burn away the dust, but as was mentioned above, this fate is far from established at this time.

The simpler molecules, such as OH, formaldehyde, and carbon monoxide, are found not only in the dense clouds of dust in which star formation is now proceeding, but are found spread throughout the Milky Way galaxy. While these molecules are still associated with dust, they are also found in the less dusty regions that form the more typical regions between the stars.

Which molecules are likely to be discovered next? While no one knows the answers ahead of time, there is a logical sequence of molecules that is expected to be found and systematic searches for these are under way. For example, the discovery of ethyl alcohol, an important link in the chain of molecule formation, was announced in 1974. More molecules are continuously being sought. A tremendous sense of competition between astronomers pervades this area of research — a competition that belies the usual picture of how scientists make progress in a orderly, carefully thought-out way!

Knowing how molecules form in space is still a problem since we can easily see that

The Horsehead nebula in Orion in which several species of interstellar molecules are found. The nebula is actually a dense dust cloud that totally obscures light from beyond it.

The 12-meter diameter radio telescope of the National Radio Astronomy Observatory located atop Kitt Peak near Tucson, Arizona. It has been extensively used in searches for interstellar molecules.

this process in nearly empty space is not going to be the same as in a test tube in the laboratory. In the lab we can mix gases and shine ultraviolet light on the mixture or flash electrical sparks through it and watch the generation of more complex molecules proceed quite rapidly. In space, however, there are so few atoms around (perhaps only hundreds or thousands in every cubic centimeter) that they hardly ever bump into each other and stick together to form a molecule. It is thought likely that some molecules in space form because they stick together after initially colliding with dust particles (whose nature is unknown). The dust particles collect lots of atoms (and simple molecules) that then combine on their surfaces to form more complicated ones. Then the more complicated molecules must be driven off the dust particle again so that they may wander freely through space, allowing us to pick up their radiations (light, infrared, or radio signals). If they remained stuck to the dust we would never be able to detect them.

Another process of molecule formation involves the effects of cosmic-ray particles that strip electrons out of orbit about various atoms. The resultant ions formed readily combine with other ions and ultimately, when the electrons are captured again, the larger molecules are built up. These processes are all part of the new science of astro-chemistry (which has found chemists becoming astronomers and vice versa), and it is one of the most exciting branches of research today.

In order to emphasize the importance of the molecular discoveries now being made out in space, we have indicated in Table 12.1 those molecules that are also produced in laboratory experiments intended to simulate

TABLE 12.1 List of Molecules Found in Interstellar Space

Molecule	Common Name	Year Discovered
CH	—	1937
CN	Cyanogen	1940
CH+	—	1941
OH	Hydroxyl	1963
H_2O	Water	1968
NH_3	Ammonia	1968
* H_2CO	Formaldehyde	1969
* CO	Carbon Monoxide	1970
H_2	Hydrogen molecule	1970
* HCN	Hydrogen Cyanide	1970
—	X-Ogen	1970
* HC_3N	Cyanoacetylene	1970
* CH_3OH	Methyl Alcohol	1970
* HCOOH	Formic Acid	1970
CS	Carbon Monosulphide	1971
* NH_2COH	Formamide	1971
SiO	Silicon Monoxide	1971
* CH_3CN	Acetonitrile	1971
OCS	Carbonyl Sulphide	1971
* HNCO	Isocynanic Acid	1971
HNC	Hydrogen Isocyanide	1971
CH_3CCH	Methylacetylene	1971
* CH_3COH	Acetaldehyde	1971
H_2CS	Thioformaldehyde	1972
H_2S	Hydrogen Sulphide	1972
CH_2NH	Methanimine	1972
SO	Sulphur Monoxide	1973
HD	—	1973
DCN	—	1973
$(CH_3)_2O$	Dimethyl Ether	1974
CH_3NH_2	Methylamine	1974
CCH	Acetyline radical	1974
N_2H+	—	1974
SiS	Silicon Sulphide	1974
CH_3CH_2OH	Ethyl Alcohol	1974
HDO	Heavy Water	1974
CH_2CHCN	Vinyl Cyanide	1975
NS	Nitrogen Sulphide	1975
NH_2CN	Cyanimide	1975
$HCOOCH_3$	Methyl Formate	1975
SO_2	Sulphur Dioxide	1975
HC_5N	Cyano-diacetylane	1976

* These molecules produced in laboratory experiments in pre-life chemistry.

conditions on this planet when the precursors to life were being formed.

The big question still concerns when the first amino acid will be found in space. Amino acids have already been found in meteorites, but their discovery in the farthest reaches of space will put speculation about the occurrence and nature of life elsewhere in the universe on a more solid scientific footing.

SECTION IV
STARS

Stars exist by the billions in our Milky Way. Where are they, how large are they, how were they born, and how will they die? These questions will be examined here. The nebulae produced as some stars throw off shells of matter in the last years of their lives are magnificent objects to view from our vantage point on earth. The story of stellar evolution, as a star grows out of a cloud of gas and dust, only to die and return some of its matter back to the space between the stars, will be traced in this section.

CHAPTER 13
HOW FAR IS UP?

We take it for granted that the stars above us are far away, but just how far away are they, and how do we know? The problems associated with finding the distances to stars and galaxies are some of the most fundamental, most complex, and most fascinating in astronomy.

The puzzle of finding distances to astronomical objects has many pieces. The first important piece involves finding the distances to objects like the planets and the sun. This is done in modern-day astronomy by forming a model of what the solar system is like and how the planets move about the sun. The planets orbit the sun according to the laws of planetary motion discovered by Johannes Kepler in the 17th century. Using these laws we get the relative scale of the solar system and if we can find the distance between any two bodies in it, for example, the distance of Venus from the earth, then we know the scale of the whole solar system.

The distance of Venus from the earth is found by bouncing a radio signal off Venus and timing how long it takes for the echo to reach us. This is called planetary radar and such experiments are part of radar astronomy. Because we know the speed at which the radar signal was traveling (equal to the speed of light) we can calculate the distance the signal traveled in the time it took to receive the echo. From this the scale of the solar system follows, and then we can fly rockets and spacecraft to the moon, or Mars, or Venus, since we know where they are in space. One also finds the sun-earth distance (called the Astronomical Unit = 150 million kilometers) in this way.

To get the distances to the nearer stars we make use of a phenomenon called parallax. Imagine looking at a telephone pole about 100 meters away, located, say, 10 meters in front of a house. Now, if we move to the side by a step or two, the pole will appear to move a little with respect to the house. We will then see it in front of a different part of the house. Now look at a much more distant telephone pole in front of another house and move a step or two to the side. This time the pole will appear to move only slightly across the front of the house. If we were to look at a pole and a house so far away that we needed binoculars to see them and we did the same experiment, we would find that the pole would not appear to move across the front of the house at all.

The same effect can be observed for stars. Because the earth is in an orbit about the sun, the earth moves the equivalent of the step to the side, and back again, once every year. Therefore, the nearer stars, as seen against the backdrop of the very distant stars, appear to move back and forth across the sky during the year. This is called parallax. The effect is very small, but can be measured using precision instruments and telescopes. By simple geometry the astronomer can then calculate the distance to the stars that show this effect. By observing the parallax of the stars, distances out to about 500 light-years can be measured. However, 500 light-years is only a very small distance by astronomical standards.

If it were not for the existence of one set of variable stars, the Cepheid variables, we would have a much harder time knowing where things are in space. These stars vary in brightness in a regular way and are used to give distances to groups of stars in which they are located. We could do this even when it was not understood why a Cepheid variable star varies at all! Cepheid variables (named after one bright example of its class, Delta Cephei), vary over periods of days and they were used in the early 20th century to determine the distance to the clouds of Magellan (nearby galaxies) and the Andromeda galaxy, showing that both were well outside our own Milky Way galaxy. Subsequently (and this occurred in the 1920's), it was also proved beyond doubt that there were other galaxies, billions of them, in our universe, since for the first time a distance indicator was available that showed these galaxies were very far away.

The story of the Cepheids began in 1912 when the astronomer Henrietta Leavitt studied the Cepheids in the Magellanic clouds. Leavitt discovered that there appeared to be a direct relationship between the star's brightness and the period of change in its brightness. This had never previously been noted and appeared to present a clue to getting a distance to the stars. Finding the distance works as follows: All the Cepheid variables in the Magellanic clouds, which are two small galaxies well outside our own galaxy, are located about the same distance from us. They are all very far away; although one Cepheid might lie on this side of the Magellanic clouds and another might lie on the other side, the distance to the clouds (170,000 light-years) is so large compared to their width (30,000 light-years) that such an effect does not really matter. It is like living in central New York city and saying that everyone in Los Angeles lives the same distance from you, while someone living in the near side of the city (Staten Island) might be living

HOW FAR IS UP? 87

Our neighbor "twin" galaxy, M 31, in the constellation of Andromeda. This galaxy can be seen with the naked eye on a clear night in the northern skies just below the constellation of Cassiopeia. Orbiting M 31 are two elliptical galaxies seen in this photo as well. The spiral arms in M 31 are dramatically evident as outlined by the lanes of dust within them. This galaxy is located about 2 million light-years from us.

closer to you than someone on the far side (Queens). Since Los Angeles is about 5,000 kilometers away, another kilometer here or there does not account for much over such distances and all of Los Angeles is approximately the same distance from New York.

Before we can understand the next step it should be noted that Henrietta Leavitt had observed the apparent brightness of the Cepheids; that is, she had measured the amount of light that reached the telescope after the journey from the Magellanic clouds. If she could find out how far the clouds were she would be able to find out how bright the Cepheids actually were. That is, she would know the absolute, or true, brightness of the stars. Of course, the distance to the Magellanic clouds was not known, except they did appear to be very far away. The point now is that if one somehow knows the absolute brightness of a star and then measures its apparent brightness, one can estimate the distance the light must have traveled in order for it to have become dimmed to the level measured. (We know that light intensity falls off as the distance increases.) This is true for all stars, of course, but the important point is that since

This photo of the outer parts of the Andromeda galaxy, M 31, clearly shows the stars and dust in the spiral arms. Most of the stars scattered uniformly throughout the photo are stars in our own galaxy.

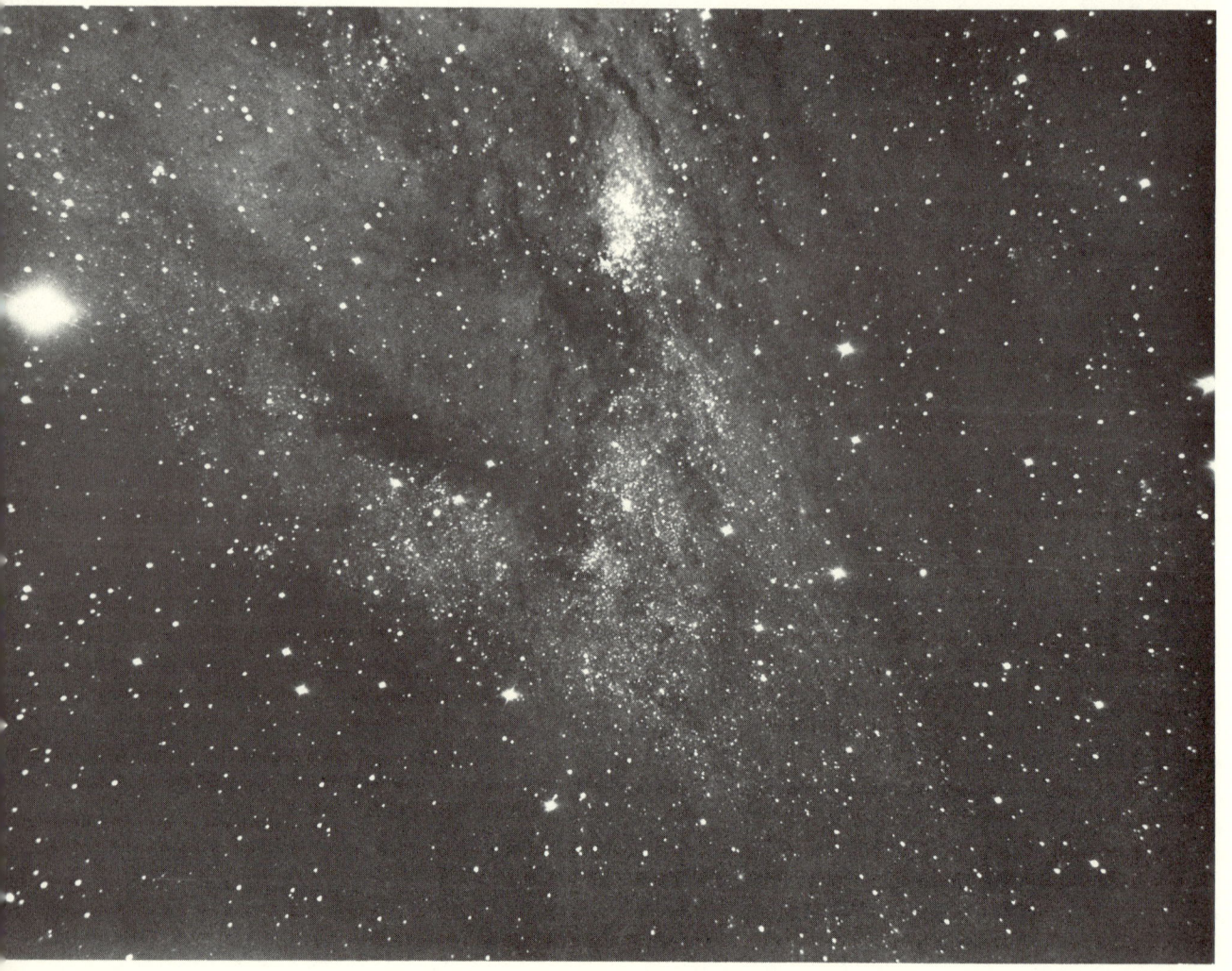

the brightness of the Cepheids seems to depend on the period in a simple way, one should be able to get the absolute brightness of any Cepheid once its period is measured. Unfortunately, there was no way to do this for the Magellanic cloud Cepheids, since the absolute brightness for any of them was unknown.

However, if somehow it was possible to measure the absolute brightness of any Cepheid variable anywhere in our galaxy, using some other method, then we would know that a period of, say, three days corresponds to a certain absolute brightness for that type of Cepheid. This knowledge could then be applied to the Magellanic cloud Cepheids. We would see how bright a Cepheid of three day period appeared to be in the clouds and knowing how bright it should be, we could obtain its distance. (The true brightness of any other Cepheid with a different period could also be automatically found because of the law relating these two properties, brightness and period.) The period-brightness relationship had never been noticed for other Cepheids nearer the sun (in our galaxy) because they were spread over all sorts of distances and this affected their apparent brightnesses in random ways. Only while studying a sample of these stars at the same (large) distance from us was the relationship recognized.

The exciting aspect of the phenomenon found by Leavitt enabled her to prove conclusively at last that the Magellanic clouds were so far away that they had to be other galaxies. Before the 1920's the existence of other galaxies in our universe had been suspected but not proved.

To get the distance and, hence, absolute brightness of at least one Cepheid to permit the calibration of the so-called period-brightness law, an interesting experiment for determining distance was done as far back as 1913. The method involved examining a large number of one type of star in a particular part of the sky. We first have to assume that these stars are all of similar brightness and that their average motion in a direction along the line from us to them is the same as their average motion in some other direction at right angles to it (across this line, on the plane of the sky). This would be true if their motions were random. The motions of the group of stars along the line toward us is found by measuring their velocities. This is found by examining the shift in the spectral lines from the stars (measuring their velocities from the Doppler effect). For example, the measurement might tell us that in one year the average motion for the stars might be one ten-thousandths of a light-year toward us. The motion across the sky for these stars is also measured. Such a change in position of the star as seen against more distant background stars is called its proper motion. It is the actual change in position that occurs because the star is moving through space along some path. We should not confuse this with parallax, which is the motion the nearby stars appear to have due only to the earth's movement about the sun.

If we measure the proper motion and find that it is, say, one second of arc in one year, then we know that one second of arc (across the line of vision) corresponds to one ten-thousandth of a light-year (along the line toward us) for that group of stars. It is then easy to estimate the distance at which a line of a length of one ten-thousandth of a light-year would appear to cover an angle of one second of arc. The average distance to the stars can then be calculated. This method was applied to nearby Cepheid variables, and although highly uncertain, it did allow their absolute magnitudes to be approximately determined. In this way, the period-brightness could be used to find absolute magnitude for any other Cepheid simply by measuring its period. This information was then applied to the original data and the distance to the Magellanic clouds was found. After that, the distance to any galaxy containing a Cepheid variable could also be estimated.

It was then apparent that the universe was indeed full of distant star-islands (galaxies) like the Milky Way. Unfortunately, it appeared that the Milky Way was the largest of them! This was shown to be incorrect about a quarter of a century later, when it was discovered that there were two types of Cepheid variables. One type was associated with young stars and another with older stars, and their period-brightness laws were different. When this correction was incorporated and the distances sorted out, the universe changed its size and the Milky Way was no longer a freak! This whole episode took many decades to resolve and today much more goes into studying distances to stars and galaxies. However, the Cepheids did confirm the extragalactic nature of the Magellanic clouds and lead to the establishment of the first concept of how large the universe actually is.

Cepheids are the most useful tool for getting the distances to galaxies in which Cepheids can be seen, but there are, of course, countless galaxies that are far too distant for individual Cepheids to be visible. Finding distances to other galaxies depends on very different methods that involve measuring their so-called redshifts.

The light from any star or galaxy (the composite of all the starlight in the galaxy in that case) shows very distinct, dark absorption bands (or lines) at particular wavelengths. This is due to the absorption of radiation by particular atoms. Each atom has its own signature in the spectrum of radiation from the stars and galaxies. In the case of distant galaxies there are several basic atoms that are always seen to be absorbing the light from the galaxy. Important absorption lines are produced by calcium and magnesium.

In 1926 it was discovered that the light from distant galaxies showed that the well known absorption lines were shifted in wavelength toward the red part of the spectrum in virtually all galaxies. The red end of the spectrum is the longer wavelength end and this redshift is possible only if all the galaxies are moving away from us (the redshift is understood to be a Doppler shift due to motion of the object away from us). The conclusion that all galaxies were moving apart was unavoidable. In 1929, Edwin Hubble showed that the fainter galaxies (which are more distant) appeared to be moving away from us faster than the brighter, and presumably nearer, ones. After the distances to some of the nearer galaxies had been determined using Cepheid variables, he established a definite relationship between the distance to a galaxy and its redshift.

This relationship between distance and redshift became known as the Hubble law. It states that a galaxy is receding from us at 18 kilometers per second for every million light-years of distance from us. The original value of this Hubble constant (as it came to be called) was thought to be 180 kilometers per second (km/sec) per million light-years, but the most current estimates place it near 18 km/sec per million light-years.

Nowadays, to estimate the distance to a newly discovered galaxy or quasar we measure the redshift and if it is, say, 2,000 km/sec, we know that the distance must be about 111 million light-years.

At this juncture it might be pointed out that the redshift phenomenon showed that the more distant galaxies were moving away from us at greater speeds than those closer by. This can be understood if we live in a universe that is expanding in all directions. Objects that are further away are then expected to be moving away from us at greater speeds. This will be discussed further in chapter 27.

Refinements in the ways distances to stars and galaxies are measured are continually being made and the chain of information required to get the final results, such as Hubble's constant, is very complex. Only a very simplified overview has been given here.

CHAPTER 14
MEASURING STAR DIAMETERS

To the unaided eye, practically everything in the sky except the sun and moon appears like a pinpoint of light. Some of these pinpoints are planets — usually the brightest ones — and the rest are stars. But the difference between planets and stars goes far beyond a mere discordance in brightness. Any small telescope focused on Jupiter, Mars, Venus, or Saturn will instantly reveal a disk. We perceive another world whose clouds or rocks reflect the sun's light back to us. But that same telescope, when turned to the stars, shows nothing new. We still see pinpoints — they just look brighter in the telescope.

What about a larger telescope? Even the 200-inch Mt. Palomar giant still shows all stars as looking about the same. If we tried to measure their visual diameters, the bright ones would seem to be larger than the faint ones — but that is only an illusion caused by atmospheric turbulence (or "seeing") and a property of optics called diffraction. The stars are simply too far away for their diameters to be measured directly.

Since no telescope on earth is capable of revealing the tiny disk of a star, how do we do it? How can we say that Sirius is 1.8 times the diameter of the sun? Yet that statement appears in many astronomy books. How is it done?

Until recently it was educated guesswork. We know how far away Sirius is; we know its apparent brightness, and we have studied its light emissions. Using this information, plus some additional assumptions about stellar physics, we get a rough estimate of the diameter. But suppose we are wrong about some of our assumptions; how can we check them? It would be reassuring if we could actually measure a star's diameter.

Now, thanks to a remarkable new telescope, that elusive astronomical endeavor — measuring the diameters of dozens of stars — has been achieved.

The star-measuring telescope is the stellar interferometer at Narrabri in Australia. It consists of two 7-meter-wide mirrors separately mounted on "trucks" that move around a circular railroad track about 200 meters in diameters. Each of the mirrors is actually made up of 252 individual, small, hexagonal glass mirrors. During the day, the telescopes are driven into a garage straddling the track to protect them from sun and wind. The strange thing about these telescopes is that you cannot even see very clearly with them, because of the nature of their surfaces. Having 252 separate mirrors makes the whole thing fairly rough — in order to get a good image in astronomy, a highly polished, very smooth surface is necessary.

The stellar interferometer was not built for photographing objects in space; it was built only to measure the sizes of stars. The two large mirrors are used as "light buckets," collecting as much light as possible and focusing it to one point; each of the 252 small mirrors in each telescope does this separately. The light is then focused onto a photomultiplier — a device for converting the light to an electrical signal — that in turn can be amplified (electronically magnified) and processed by a computer.

The basic principle of this interferometer is this: If a light signal (a photon) from a distant star simultaneously reaches each of two telescopes located at the same distance from the star, then we say that the light received by one telescope is correlated with the light received by the other. Using computer circuits, the amount of correlation can be measured. If these measurements are made with telescopes separated by the maximum distance on the track, only the smallest stars will still be observed in correlation. Light from the larger stars will seem "confused" until the telescopes are pulled closer together. This method provides comparative diameters from which the actual diameters eventually can be derived.

The stellar interferometer at Narrabri is not limited by seeing or diffraction as single conventional telescopes are. It is only limited by how far apart the telescopes can be located so the light can still be converted to electrical signals that can be compared in a computer. There are a lot of technical difficulties associated with such an experiment; the computer is not only comparing the signals from the star, but is also comparing signals from the photomultiplier, the night sky, interference, and so forth.

R. Hanbury Brown, who spent more than ten years working on this interferometer, faced many unexpected problems in building such a complex system miles from anywhere. The scientists, in the best tradition of pioneering research, soon became adept at welding and other nonastronomical trades.

In order for the two mirrors to be precisely located with respect to the stars, the trucks had to be accurately driven and positioned — which meant that the tracks had to be very accurately leveled. The complexity of the experiment almost defies description. For example, the light from a star has to reach each of the photomultipliers at precisely the same time and then remain precisely in sync until the signals are compared in the computer. If there is a difference of more than one thousand-millionth of a second between the two paths, the experiment fails.

About twenty years have passed from the

TABLE 14.1 Measured Star Diameters[a]

Star Name	Magnitude (apparent)	Measured Diameter (seconds of arc)	Actual Diameter (sun=1)	Luminosity (sun=1)	Temperature (degrees Kelvin)	Distance (light-years)
Alpha Canis Majoris (Sirius)	−1.4	0.0059	1.8	23	13,500	8.7
Alpha Carinae (Canopus)	−0.7	0.0066	24	1500	8,500	100
Alpha Bootis (Arcturus)	−0.1	0.020	23	110	5,000	37
Alpha Lyrae (Vega)	0.0	0.0032	3.0	53	14,000	27
Beta Orionis (Rigel)	+0.1	0.0025	77	55000	16,500	900
Alpha Canis Minoris (Procyon)	+0.4	0.0055	2.1	8	7,000	11.3
Alpha Orionis (Betelgeuse)	+0.4	0.054[b]	800[b]	15000	3,000	500
Alpha Eridani (Achernar)	+0.5	0.0019	8	700	18,500	118
Alpha Aquilae (Altair)	+0.8	0.0030	1.9	10	10,000	17
Alpha Tauri (Aldebaran)	+0.9	0.020	45	400	4,500	68
Alpha Scorpii (Antares)	+0.9	0.040	400	9000	4,000	400
Alpha Virginis (Spica)	+0.9	0.0087	7	1800	25,000	220
Alpha Piscis Austrinus (Fomalhaut)	+1.2	0.0021	1.8	13	12,000	23
Beta Crucis	+1.3	0.0007	13	5500	28,000	490
Alpha Leonis (Regulus)	+1.4	0.0014	4.3	160	17,000	84
Epsilon Canis Majoris	+1.5	0.0008	16	10000	25,000	680
Gamma Orionis (Bellatrix)	+1.6	0.0007	14	3700	25,000	470
Beta Carinae	+1.7	0.0016	5	100	14,000	86
Epsilon Orionis	+1.7	0.0007	40	40000	28,000	1600

[a] Arcturus, Betelgeuse, Aldebaran, and Antares were measured in 1920 with an interferometer attached to the Mt. Wilson 100-inch telescope. The other stars were measured with the device described in this chapter.
[b] Betelgeuse varies in size and brightness irregularly over a period of years. The diameter is the maximum. The listed magnitude is the average value.

time Brown started thinking seriously about the experiment until 1974, when an article containing diameters of thirty-two stars appeared in the Astronomical Journal. Twenty years seems a long time to spend measuring the diameters of a handful of stars, but such dedication seems to be part of the advance of science.

The largest diameter measured so far by the Australian interferometer is 0.0087 seconds of arc for Spica, and the smallest is 0.0004 seconds of arc for Zeta Puppis and several other stars. These measurements are the apparent diameters. To find the true diameter the distance has to be included in the calculations. The nearer stars often appear larger for that reason (just like Mars looks larger than Neptune, when in reality it is the other way around). The actual diameters are included in Table 14.1.

Measuring diameters this small is not a matter of going to the telescope, pointing at the stars and taking a reading. It requires up to fifty hours of continuously tracking one star at a time, recording all the data, and then processing them for an answer. That means that ten nights might be spent tracking just one star and storing the data until they are all available for processing.

Basically the measured diameters of the stars agree well with those calculated years ago. For example, Vega has an apparent diameter of 0.003 seconds of arc — just what was expected from theory.

Before the new data became available, less than a dozen diameter measurements (made early in this century with the Michelson interferometer by F. G. Pease) were available. Some of these are included in the table to give a broader cross section of star types. Unfortunately the Narrabri interferometer cannot measure many more stars than it already has, since most are too faint for the present "light buckets" (mirrors) to collect enough light to attempt the measurement. It is possible to build a larger interferometer based on the same principles, but funds are not yet available.

Does this mean that no more stellar diameters will be measured in the near future? No, because strange as it may seem, the moon can be used to measure star diameters. When a star disappears behind the moon, called an occultation, the way the light signal disappears can provide information allowing an estimate of the stellar diameter. However, this seemingly simple experiment is very difficult in practice. There are many irregularities on the moon's surface that make for trouble. Twinkling in the atmosphere and random signals in the telescope affect the observations. These difficulties limit the number of stars whose diameters can be measured by the lunar occultation technique to a few hundred over the next twenty to thirty years.

CHAPTER 15
FROM DUST TO DUST

Nothing seems more permanent than the stars. Year after year the same stars, in the same constellations, shine as faithfully as the sun. With few exceptions, the stars you see tonight have been in the same locations for thousands of years — some for millions of years.

When the dinosaurs roamed earth, the skies were somewhat different. There were probably more bright stars then, many of them closer together. But even then there were a lot of stars that have not changed since. A star lasts a long time before its supply of fuel is consumed and it dies quietly or explodes violently, depending only on its mass (or so astronomers now believe).

Just like us, all stars are born, grow older, and die. And just like us, what happens to the star during its lifetime determines what the next generation of stars will be like. As stars grow and die, they spread newly processed material to the interstellar medium (the space between the stars) — material which will be used to form future generations of stars.

There was a time, before any stars were born, when the universe was mainly hydrogen gas with some helium and very little else. That was perhaps 15 billion years ago. Somehow stars were born and started to shine; the hydrogen gas was "cooked" to helium, and then to heavier elements as the stars grew hotter. In those early days of the universe there were probably no planets like earth because there was so little of the heavy elements — no iron, no silicon, and little oxygen — and thus no solid planets.

But as time passed, these heavy elements were being brewed inside the stars and wafted or blasted out into space to create the dust clouds that we now see among the stars. These are the dust clouds that contain

The bright stars in the Pleiades cluster outshine the hundred or so other members in this photo. These stars are embedded in dust and gas that glows in the same way bright street lamps illuminate haze surrounding them.

dozens, perhaps hundreds, of different molecules. Already, forty types have been found, and yet we still do not know what the dust itself is made of (see chapter 12). We do know that stars are still being born. Yet there is no population explosion, because as stars are born, others die. There is a finite quantity of matter available in a galaxy, so the process will ultimately end.

Today more time and effort than ever before is being spent on trying to understand exactly every step in the evolution of stars. Let's begin with the clouds of gas and dust and see what happens to them.

The modern astrophysicist attacks this problem by making a "model" of a cloud that can be fed into a computer. The computer is given various equations — the known laws of nature — and is then run until it "gives birth" to a star. It all sounds rather magical, but the model is really only a mathematical formulation of the conditions that are thought to exist in the cloud before the star forms — the protostellar cloud as it is called. Since we cannot watch a cloud for thousands of years to see what happens to it, this computer model is our only alternative.

The computer is fed both the physical parameters that describe the cloud and the equations that explain, according to presently known laws of the universe, the way in which matter will behave as time progresses. For example, we must consider how gravity balances rotation in a protostar, how heat generated inside the star can escape (whether by radiation or convection), and how the temperature depends on size.

Depending on the state of the star at any time, the equations will predict what the next state will be. What will the temperature become if it is this hot now? How large will the star be next if it is now so large and so hot and so old? Such questions are fed into a computer and, as the computer ticks away, a second of computer time might be 1,000 years of star time. Finally, the researcher looks at the numbers that the computer prints out and compares them with the properties of real stars.

Throughout the galaxy we have a sample of stars of all ages. In a random sample of 1,000 stars we will find that some will be very young (maybe 100,000 years old), and others will be very old (perhaps 10 billion years). We measure their properties as best we can to determine their ages. Then we compare these quantities with those that were generated in the computer. We do not have to wait a billion years to watch the real star evolve so that we can compare it with our model. Nature has seen to that. We merely look out at a large sample of stars above us and then find all sorts of them already there — young stars, old stars, hot stars, cool stars.

It is just like taking a census here on earth. If we want to find out about the evolution of humans we do not have to sit around and wait for eighty years. We can instead take a sample of one small city. We need to look at who is living there at any one instant: Count the number of babies, children, and adults; weigh them, measure their heights, and find their ages. We will then have an instant pictorial summary of just how humans evolve. They start off young and tiny, and rapidly grow to maturity (spending about fifty years in that state); their sizes and weights change relatively little; then they die. Most of our sample of humans would be classified as adults, which obviously means that humans spend most of their lives as adults.

By looking at a group of stars all born at the same time (a cluster or an association, as such a group is called), we get a snapshot of the way stars evolve. We can see how long they stay in any given state. If most stars in our sample are a certain temperature and brightness, that means that a typical star will spend most of its life that way.

The existence of clusters and associations means that we have stars that have grown, or are still growing, out of one cloud of primeval gas. We can compare the properties of cluster stars with a larger sample of stars spread randomly about us — the so-called field stars — and we will find that they, too, show a similar distribution of ages, temperatures, masses, and brightnesses. Because we are seeing the state of individual stars as they are now, this can be interpreted as a profile of the way stars in general evolve.

The careful reader might say that some of the stars are farther away than the others in our sample so that the light of those stars has taken longer to get to us. Those stars are, therefore, seen at a different time from the others, and we no longer have an instantaneous snapshot of the sample. Is our analogy of taking an instant census of a town incorrect? No, because the stars are very close to us relative to their lifetimes. In our sample most stars are at least hundreds of thousands of years old, but their distances are only thousands of light-years at most. A thousand years of time is an instant in the evolution of a star, so we need not worry about the light travel time.

Let's trace our investigation of stellar evolution to a starting point, somewhere in interstellar space. Here, hydrogen gas — making up 10 percent of the mass of the Milky Way galaxy — drifts and swirls between the stars in enormous, dust-filled cloudlike structures. The study of these clouds of hydrogen gas lies entirely within the field of radio astronomy. Prior to 1951 they were unobserved; in that year radio signals from the hydrogen atoms were discovered at a wavelength of 21 cm. By tuning a receiver hooked to a radio telescope to this wavelength, the signature of hydrogen can be easily detected.

Radio telescope observations of the Milky Way have revealed that hydrogen is everywhere in the galaxy. Analysis of the signals gives the distance of the gas, the shape of the clouds, their temperature, mass, and their relationship to other objects in the galaxy. The clouds range in temperature from a few degrees to thousands of degrees above absolute zero. The coldest ones are those in which much of the interstellar dust is found; it is in these clouds that star formation is likely to be taking place. But how does a cloud of hydrogen gas give birth to a star?

Imagine a gigantic cloud, hundreds of light-years in diameter, but so dispersed that you would not be aware of being inside it (assuming for a moment that we could travel to the cloud and look out a window). Over a lengthy period, as a result of pure chance, more matter gathers in one region than in some other region of the cloud. This small condensation tends to attract surrounding matter to it, and our original cloud of gas develops irregularities that grow in size. Eventually gravitational forces acting on such condensations will collect more matter; the subsequent increase in gravity will cause the blob to contract and grow

100 STARS

The dark dust clouds known as Barnard 68 and Barnard 72, seen projected against the great star cloud in Sagittarius. It is within such clouds that complex interstellar molecules are found.

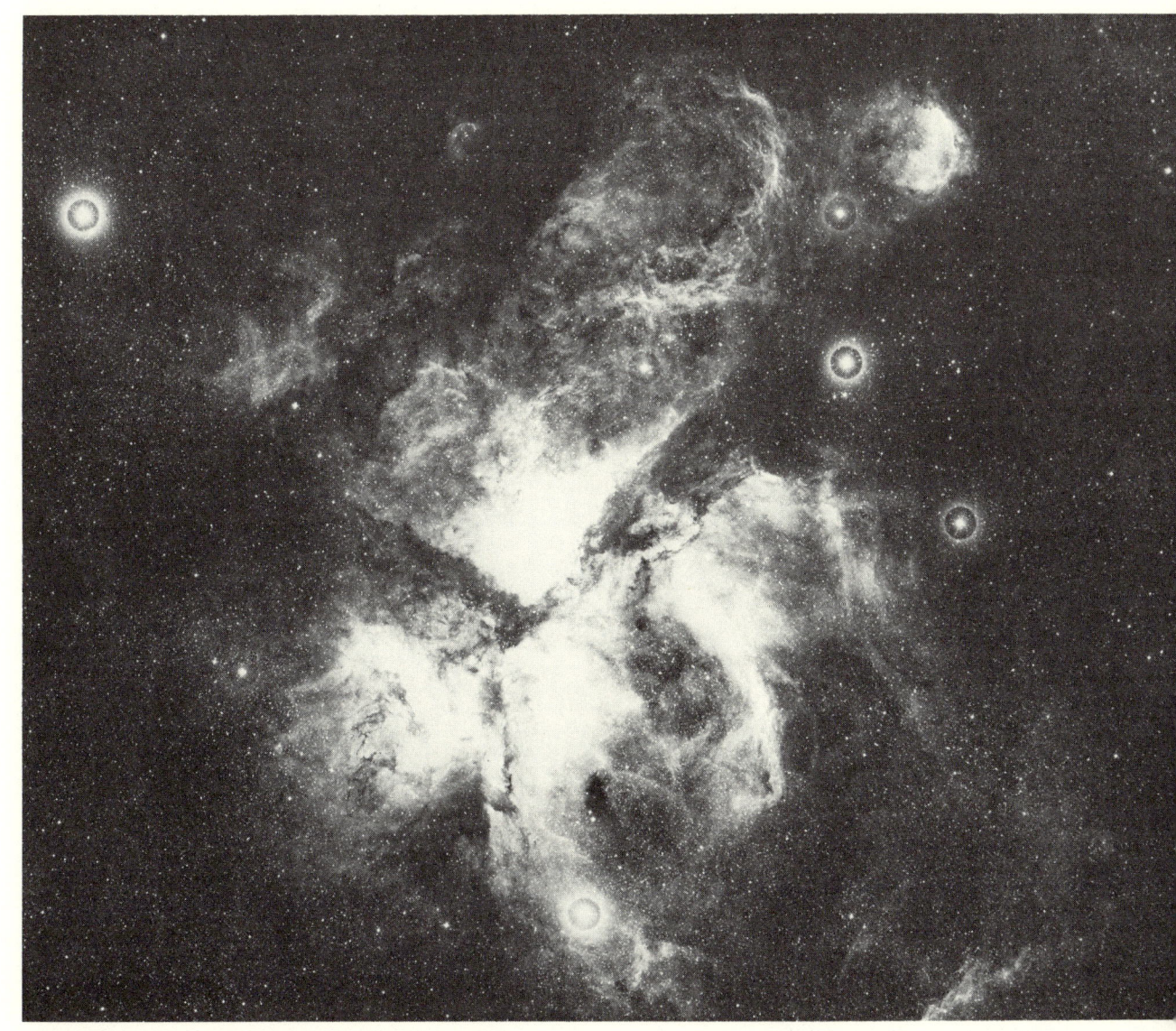

The Eta Carina nebula. This is a beautiful example of an emission nebula in which a cloud of gas has been heated by hot young stars so that the gas emits its own light energy. Clouds of dust produce dark markings across the front of the nebula.

smaller. Soon our original nebula consists of many smaller clouds of matter, many of which are physically growing smaller while all the time gathering more matter into them. Some collide with others or decompose because there is too much heat inside them to balance the effect of gravity. (Heat inside any object — whether a cloud or a star — tends to cause the object to expand; gravity acts in the opposite way, causing it to contract.)

Now imagine we are watching just one of these condensations in our original large cloud. As it collects more matter, it contracts further, and at the same time the atoms inside collide more often. The rate of such collisions is a measure of the temperature of the object: An increase in collisions due to an increase in the amount of gas in a given volume of space means an increase in temperature of that gas. This additional heat can be radiated away into space tending to cool the cloud; if it is not radiated away quickly then the heat will actually prevent the cloud from contracting any further. Should the condensing cloud heat up too much it will disperse.

The hydrogen clouds studied by astronomers are often well defined and fairly small (10 light-years across), and must accordingly have lost enough heat so that gravity is still in control of their further shrinkage. Radio measurements can give the temperature of — and the amount of matter in — the clouds. Therefore, we can calculate how strong the effect of gravity is and, because of its temperature, how much force is acting to disrupt the cloud. The gravitational force should be greater than the outward force produced by the heat in the cloud — otherwise the cloud would not exist. Strangely enough, that is hardly ever the case.

There are virtually no clouds observed in which gravity is strong enough to overwhelm the temperature effects based on the measurements that can be made at present. Is the picture suggested above all wrong? No, we do not think so. But then we do not know what the solution is either. Basically there does not appear to be enough matter in any of the hydrogen clouds in the Milky Way that would allow them to contract and be stable. Apparently our attempt to explain the first stages in star evolution has failed. But then why are all those clouds there, and why do we see stars?

The answer requires considerable speculation about conditions in space. After all, we have only been talking about the hydrogen gas in these clouds, gas that is observed by radio telescopes at 21 cm wavelength. Hydrogen is just one component of interstellar matter. We have not included molecules, dust, black holes, and who knows what else that might exist within the clouds.

There may be a way out of this dilemma. The dust and the molecules in the clouds are now known to be a small fraction of the cloud's mass, and do not help to increase the gravitational pull of the cloud. Fortunately, however, there is one component of matter that we find very hard to observe and is expected to exist in these clouds — molecular hydrogen. This molecule (two hydrogen atoms stuck together) has so far only been observed in the direction of a few stars by means of ultraviolet telescopes; we really do not yet know how much molecular hydrogen lies in typical atomic hydrogen clouds. Such a situation is tailor-made for any theoretician to work with because there are no data that could contradict any assumption made about the amount of additional matter in the clouds.

Although it appears that hydrogen clouds could not possibly exist because they do not contain enough hydrogen to produce the necessary gravitational pull, we must conclude that there probably is enough additional molecular hydrogen present that we have not yet found. If we want to be more daring, we might conjure up some black holes to account for the missing mass. This missing mass problem is one that is found in many areas of astronomy, not just in hydrogen clouds. Many star clusters and clusters of galaxies do not contain enough documented matter to prevent them from expanding outward and, thereby, ceasing to exist.

To get back to our cloud of contracting gas, we assume it must have enough matter to cause it to contract. In the meantime, it radiates away the heat that is generated as a result of contraction. Now, however, we find that many of the densest clouds (which are also the coolest ones) contain dust particles and complex molecules. Neither the origin of the dust nor the processes of molecule formation are yet totally understood. What we do know is that some of the molecules like CO (carbon monoxide) are very efficient radiators of heat, and their existence will speed up the cooling down of the cloud. This means that the cloud in which CO exists might contract fairly quickly because it rapidly loses its heat.

But this use of CO now presents a difficulty. The molecules observed in these dense clouds do indeed contain heavy atoms such as carbon and oxygen (as in CO). But these atoms had to be made in stars in the first place. Now we are saying that clouds contract to form stars because they cool down with the help of some good radiators, in the form of CO and other molecules. In that case, how did the first generation of stars — those born before there was much matter in the universe other than hydrogen gas — get formed? There was originally nothing to help cool the clouds down. Is it possible that some of the heavier atoms were made in the big bang that started it all off here in our universe? Astronomers don't think so. Then is it possible that the first generations of stars were formed much more slowly, relying on eddies of gas to concentrate the matter? Perhaps. But whatever the case, star formation early in the life of the galaxy was probably very different from the way it is now.

Setting this problem aside we will get back to our cloud, now rapidly cooling so that — while it might have taken billions of years to reach the molecule phase — it might only take millions of years to collapse into a star. In the last stages of collapse, matter falls freely into the center of the cloud.

As the cloud gets smaller and smaller, subunits within it might individually form smaller blobs that will contract independently of the others. This will depend on how much matter they contain and how fast they can lose their heat. However, we now reach an important step. The small clouds become so dense that the dust within them starts to act as a shield that prevents the heat from escaping. This means that the center of the small, contracting blob will heat up. If conditions are just right, the heating might come too late to play any role in stopping further contraction; the cloud (now comparatively small but still much larger than the solar system) will get hotter as it continues to shrink. When it reaches a temperature of 1,000 degrees Kelvin or so above absolute zero, it will radiate so much

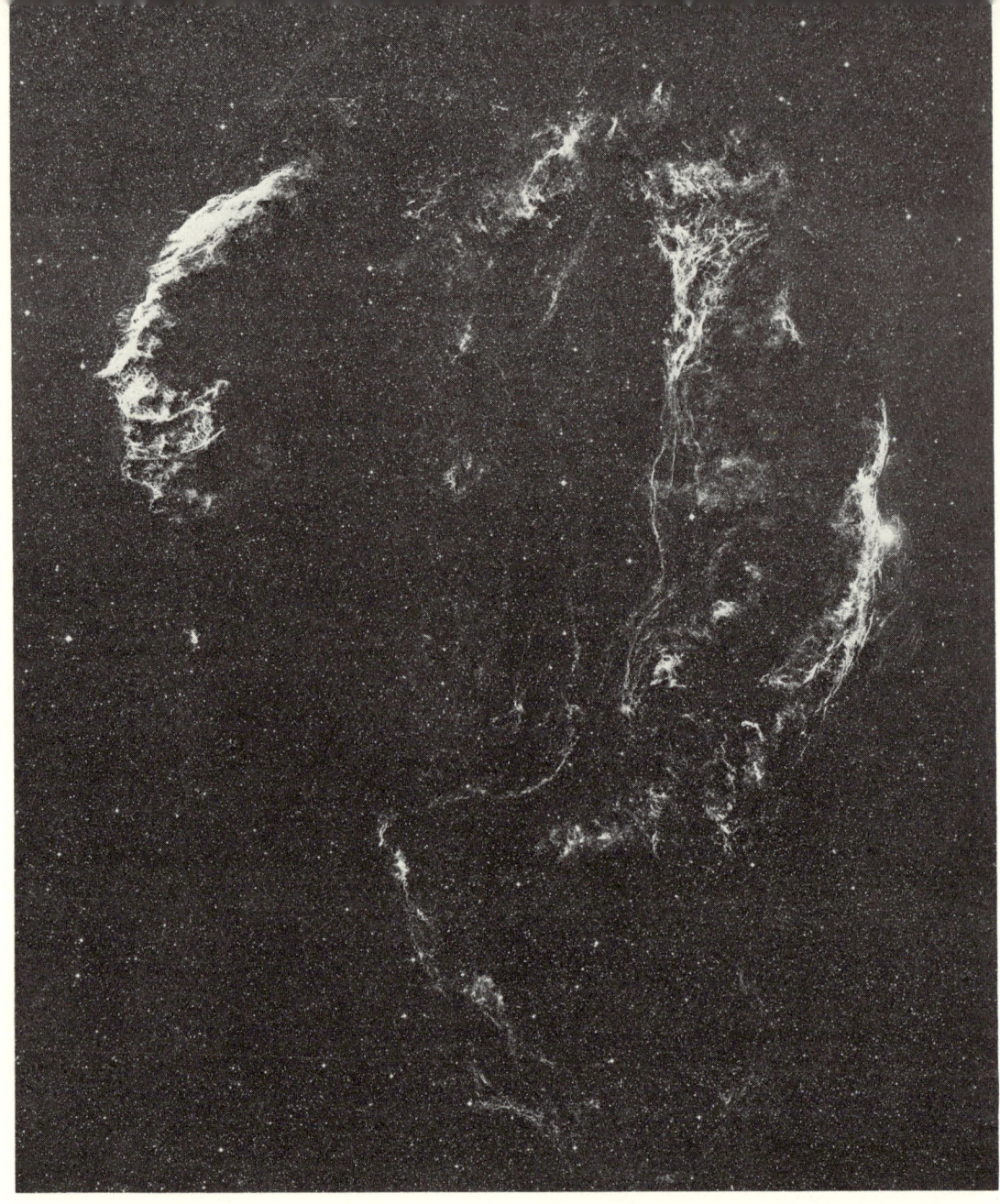

The remains of a star that exploded some time ago (several tens of thousands of years) in the constellation of Cygnus. This is an example of a supernova remnant — the shell of matter thrown out from an exploding star. The structure of this shell is partly determined by the existence, or lack of, surrounding matter to hamper its free expansion. Absence of surrounding matter probably explains why the lower left section has been blown farther outward.

heat that it is best observable from earth as an infrared source. This new phase is called a protostar.

The process continues, the object getting hotter and smaller, until at some time the temperature at the core reaches 15 million degrees Kelvin. At this temperature, nuclear reactions that "burn" the hydrogen to form helium begin. Then, as if to resist any further evolution, the star starts to oscillate wildly in brightness for a few thousand years. Only when it has settled down as a relatively steady nuclear burner can it rightly be called a star.

One of the problems in understanding how we get from a cloud of gas to a star concerns the rotation of the cloud itself. Any cloud will have a small amount of rotational motion in it. If the cloud contracts, the laws of physics say that it should spin faster. This is the well-known effect that allows a skater to spin faster when he or she pulls in his or her arms while spinning on the ice.

The simplest calculations for star formation suggest that all stars should be spinning very, very fast as a result of their enormous contraction from cloud to star; but they do not do so. Why not? The answer is far from known at present, but several possibilities exist. In order to avoid this speeding up of the star's spin, the star must rid itself of material. Various mechanisms might operate to facilitate this. For example, as the young star is getting hotter and hotter, the enormous heat being generated might cause much of the surface material to be blown away. This would allow the star to slow down its spin by carrying off the momentum in this stellar wind. Alternatively, the high spin might cause a disk of matter to be drawn out from the star — a disk that might ultimately form planets. Whatever the method whereby the star loses its spin, stars do form and they do not spin themselves into instability.

Another effect that acts to stop a cloud contracting to the size of a star concerns magnetic fields in space (see chapter 9). These fields are everywhere in the Milky Way and they control the motions of hydrogen clouds and possibly their shapes and sizes as well. Although hydrogen gas finds it very hard to move across magnetic field lines, it can easily flow along them. This means that if a cloud of gas is contracting to form a star, and there is a field present, the contraction is inhibited across the field but can occur freely along the field. This would imply that the cloud would develop an elongated shape at right angles to the field lines. Based on my own observations, I have found that this is not so and that most hydrogen clouds are elongated along the field lines! This makes nonsense of what was said above, but here again we are far from understanding all the details of how clouds actually become stars; this is one more detail that still needs to be worked out.

A more worrisome detail is that the magnetic field in a cloud increases as the cloud contracts. The field lines actually "stick" to the hydrogen gas; as you bring the gas closer together the field gets stronger. From observations of magnetic fields in clouds of different densities, we can predict that by the time the gas has contracted to become a star, the field should be millions of times stronger than the fields actually observed in stars. Somehow the field leaked away into space; otherwise the star simply could not have contracted. No one really understands how the fields leak out of the protostar. Again, a detailed explanation is still needed.

Returning to our story, we had a cloud that contracted and became the size of the solar system. It started to heat up and at some critical time — when the internal temperatures were several million degrees — the star itself started to shine. Can we be sure that any of this is correct in view of the problems mentioned above? The answer is that we probably can, since we do actually see objects that appear to be at various points along the evolutionary track. We see the gas clouds and we see very dense, compact dust clouds called globules. These globules are so dense that no light penetrates them. However, radio signals do and these signals show many complex molecules inside some of the densest dust clouds. In addition, some of the dust clouds contain infrared objects thought to be the protostars themselves. These protostars have not yet heated up to stellar temperatures, and are only several thousand degrees at most — some are as cool as 600 degrees Kelvin. Their radiation is primarily in the infrared part of the spectrum and they are observed in and around many emission nebulae, mostly inside dust clouds. Good examples are the infrared stars behind the Orion nebula (chapter 11).

In some dust clouds, the dust is so thick that even the infrared radiation cannot escape — but radio signals can still penetrate. In these cases, the radio emission from the OH (hydroxyl) molecule within the clouds is extremely strong and the intensity can only be explained if there is amplification of the signal in the cloud itself. The theories that best explain this amplification (by the maser process) require the existence of strong infrared radiators such as the protostars that we believe are obscured from our vision.

When the star first starts to shine, the heat will drive the dust and gas cloud outward and destroy much of the complex material in the cloud. The star (or stars) burns a hole in the cloud so that it suddenly becomes visible, as if a curtain has been pulled open. Such stars have in fact been seen to appear. During this century, at least two new stars (FU Orionis and V 1057 Cygni) were discovered where none were visible before. In both cases the stars were associated with small dust clouds.

After the star starts to shine, it is what we call a T-Tauri star — an irregular variable star that appears to be fighting the fact that it must soon start to shine with a steady glow. It seems certain that this oscillating stage is the one just prior to what is considered a normal star. T-Tauri stars are always embedded in nebular material where star formation is believed to be occurring. They appear to be either ejecting a lot of matter or spinning very rapidly. In either case, they are probably going to lose material in sufficient quantities so that they will ultimately be able to rotate more slowly as a normal star.

Just what is a normal star? Stars do not come in random sizes, temperatures, and brightnesses. There are certain limits, and all stars seem to fall within them. The discovery that the stars are nicely organized in this way was made early in this century by the astronomers E. Hertzsprung and H. N. Russell. They measured the temperatures of many stars (found from spectroscopic examination of their light) as well as their actual luminosities (for which distance information was needed). They found that if these properties are plotted on a graph (now called the Hertzsprung-Russell or H-R diagram), most stars fall along a clearly defined zone. This zone on the graph is known as the main sequence. Most stars in the galaxy, including our sun, are known to be main sequence stars.

Computer models of the evolution of stars have been able to account for the presence of the main sequence stars and also for the way the stars evolve onto and off the main sequence. Since most of the stars in the

The Cone nebula, situated on the outer edge of an emission nebula, clearly shows the presence of large dust clouds swirling between the stars.

galaxy lie on the main sequence, we can conclude that the majority of stars spend most of their lives as main sequence objects. Those stars that lie off the main sequence are all fascinating objects in their own right. The T-Tauri stars are located just off the main sequence, but as they evolve toward normalcy their temperature and luminosity change so that they move onto the main sequence.

For billions of years, a typical star shines steadily as a main sequence star. A star spends most of its life at one location on the main sequence because its basic properties remain largely unchanged. However, later in life its luminosity and temperature change as does its position on the H-R diagram, and it ceases to be a normal main sequence star. At this transition period, most of the hydrogen at the center of the star is converted to helium. Now nuclear burning continues in a shell around this core. The fire in the core effectively dies out. However, this means that the core is cooler and will contract, which in turn means that it will again heat up. It gets so hot that the nuclear burning in the shell is accelerated. To a distant observer the star appears to be much increased in brightness at this point.

The other consequence of these changes in luminosity and temperature is a ballooning in size. Because the amount of energy emitted by the star does not increase in proportion to its increase in size, the surface temperature drops and it changes color from yellow (for a solar type star) to red. The star becomes what is known as a red giant, radiating hundreds of times more light than when it was a normal star.

According to computer calculations, the red giant's core reaches a temperature of 100 million degrees Kelvin, igniting the helium and forming heavier elements such as carbon. The core of the star is now incredibly dense. The sudden heating produced by the burning helium happens so fast (a few hours) that a minor explosion erupts (minor by comparison to some other explosions a star can undergo). Most of the energy from this explosion is absorbed in the outer layers of the star that are still burning hydrogen.

After the helium-burning explosion (called the helium flash), the core of the star cools down and, as a consequence, the whole star cools down and once again starts to contract. Its life as a red giant is over. By contracting for the next 10,000 years, the center gets hotter all over again and the temperature gets high enough to trigger further helium burning. Now, however, there is no explosion in the core, and the star gets most of its energy from the burning helium. It continues to do this in a stable manner, although the hydrogen-burning shell is now running out of hydrogen and the helium burning in the core is spreading throughout the star. This leaves carbon (the "ash" from helium burning) as the main constituent at the center of the star. The helium, in turn, now starts to burn in a shell around the carbon core. The same effects that happened before start once again. The core of the star stops burning helium; it cools, contracts, heats up again, and tries to explode — but this time the process is more complicated. The star starts expanding toward another red giant phase but does so very rapidly (taking a few million years this time compared with hundreds of millions of years the first time). We are now reaching the last stages of the star's life.

At this stage, different stars suffer different deaths. Many will not reach this second red giant phase at all. Many will die in explosive events—some minor, others totally catastrophic to the stars and all living things within tens of light-years of the explosion. Let us consider the fate of the various types of stars.

The smallest stars — less than half the sun's mass — might never become hot enough at the center to trigger the helium- or carbon-burning phases. They simply pass quietly away, fading out and eventually becoming cold dark objects called black dwarfs. Others like the sun — or perhaps twice as large — undergo another end. For these stars, the enormous amounts of heat being generated in the core near the end of the star's life have difficulty escaping and become absorbed into the outer layers of the star that, in turn, get pushed outward. If conditions are right, this outer shell continues to expand out into space for thousands of years. From a distance the star seems to be surrounded by a cloud of smoke. The Ring nebula in Lyra is such an object.

The shell of the star that becomes a so-called planetary nebula is cool, but the regions just under this shell are much hotter (100,000 degrees Kelvin compared to 4,000 degrees Kelvin). As the shell expands away, the inner parts of the star are revealed and we find a hot star where there was a cool one before. The central star of the planetary nebula has in fact been stripped naked of its outer layer of gas. Decline now sets in as the star is running out of fuel. The central regions cool and gravity takes control — compressing the star to about the size of earth. At this size, the crushing force of gravitational collapse is finally stopped by the outward pressure of highly

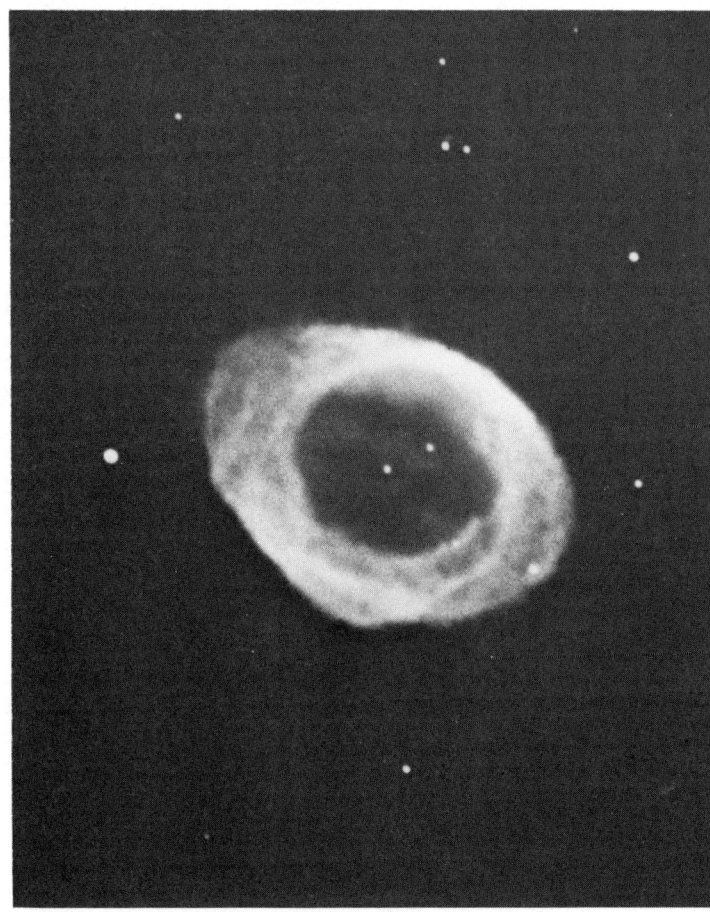

The Ring nebula, the beautiful planetary nebula in Lyra. Resembling a smoke ring, the nebula was blown out of the bright star at its center.

compressed atoms. The star is now a white dwarf, with every cubic inch containing 10 tons of matter. At an agonizingly slow rate the white dwarf radiates its last energy into space for millions of years. Finally its storehouse of gravitational energy is exhausted and it ceases to shine, becoming a black dwarf.

More dramatic — and very relevant to the existence of the planet earth as we know it — is the death of a larger star (say about ten times the mass of the sun). The interior of such a star also collapses, but now there is so much matter falling inward that the core heats up to an incredible 600 million degrees. At this temperature, carbon — the end product in a solar-sized star — is consumed by nuclear processes, forming magnesium and other heavier elements. The collapse is temporarily stopped while the carbon is burning (because of the outward pressure of its heat), but finally even the carbon is used up.

Once again, the core runs out of fuel (carbon in this case), so the heat generation stops and the collapse resumes. The temperature now increases even more as other elements are cooked up in the nuclear furnace. But when iron starts to be formed, a critical step is reached. Now conditions are suddenly very different. Instead of generating heat in the process of making elements heavier than iron (as it has in producing all the elements up to this point, iron included), the star has to use its available heat supply in producing the heavier elements. Suddenly, within only minutes, the center of the star cools down; all the available heat is swallowed in the sudden production of the elements heavier than iron. The fire in the star's center literally goes out as if suddenly extinguished, and the core collapses catastrophically.

The outer shell of the star — still merrily burning away and making heavy elements — suddenly has little below it to help hold it up against the forces of gravity. The shell promptly plunges in toward the iron core. The implosion is rapidly followed by an explosion, because the shell bounces back from this collapse (much like a spring), and it is thrown out from the star at 8,000 kilometers per second. The shell travels ever outward, propelled by the enormous flash of energy. Meanwhile the core collapse generates even more heat that ignites the outer parts of the expanding shell to enormous temperatures. The exploding star now shines with the light of a billion suns and can be seen from neighboring galaxies.

If such an explosion occurred close enough to earth (perhaps 3,000 light-years away), it could be seen in the daytime. This stellar suicide is called a supernova. One supernova is expected to occur in the Milky Way galaxy every 50 to 100 years.

All the heavy elements created in this incredible sequence of events are spewed into space. Ultimately, other stars form in clouds of recycled material, recycled to space by the supernovae. Planets form again, this time made of a mixture of heavy elements built in the death throes of millions of stars eons ago. These same heavy elements are in each of the complex molecules in our bodies and in all living organisms on this planet. We are literally children of the stars.

But the story is not over yet. What happens to the core of the star that exploded — the central parts left after the shell blew off so catastrophically? In the most bizarre of all the sequences, the core continues to rapidly collapse until there is so much compressed matter that the forces between the electrons (which prevented further contraction in the white dwarf) are completely overcome. The star continues to get denser and denser. Electrons and protons are forced so close together that they combine to form neutrons. The star is now neither gaseous nor solid,

Detailed view of a section of the supernova shell in Cygnus known as the Veil nebula.

TABLE 15.1 The Fate of Stars

Mass (sun=1)	Main Sequence Type	Evolution
20 to 60	O and B supergiants	Nuclear burning continues to iron production, then supernova; core collapses to black hole
10 to 20	O and B giants	Unstable after carbon burning; supernova; core collapses to neutron star
4 to 10	B giants	Carbon burning ignites supernova; possibly no core collapse and no dense remnant
0.5 to 4	A, F, G, and K solar type	Never becomes hot enough to burn carbon; loses mass as planetary nebula or by other processes; degenerates to white dwarf
0.1 to 0.5	M dwarfs	Never becomes hot enough to burn helium; loses mass; slow shrinkage to white dwarf
Less than 0.1	Black dwarf	Never hot enough to start nuclear fires of hydrogen burning; only heat is from contraction and radioactive decay

Data for table adapted from computations by R. L. Sears, University of Michigan.

but consists entirely of neutrons. Matter as we are used to imagining it no longer exists in this object. (See Table 15.1.)

The former stellar core has become a neutron star or pulsar. Pulsars spin rapidly — many times a second — and their spin is actually the source of more energy that can be fed into the surrounding remains of the supernova explosion. From these supernova remains, fed by the pulsar, streams of deadly cosmic rays are sent into space (see chapter 17). These neutron stars or pulsars radiate pulses of light, radio, and X-ray emissions, and are detected on earth by astronomers using telescopes operating in these parts of the spectrum.

Sometimes the neutron star — a 10 or 20 kilometer ball of neutrons — is not the final destiny of the star. If the star that originally exploded was more than about twenty times the sun's mass, then the collapse of its core does not stop at the neutron star phase. Even the forces keeping the neutrons apart are overcome by the all-pervading force of gravity. A point is reached where nothing — not even light — can escape from the object; at this point we call it a black hole. A stellar collapse black hole has a diameter of only a few kilometers when it disappears from sight. But down inside the hole the matter it contains continues to shrink forever, reaching a size so small that we cannot conceive of it in any physically meaningful way. Now only mathematics can describe the object.

These stages of evolution of stars from beginning to end are just as inevitable as those that affect our lives. Just like us, stars are born and must ultimately die to give way to the next generation. Stars created out of cosmic clouds of gas and dust usually have very long lives, but all must die when their fuel runs out. Some go quietly, others with an incredibly destructive explosion. Life on any planets will also cease when the stars they orbit reach the ends of their lives. There is no escape from that.

For earth and our sun, that time is still billions of years in the future. What is more likely to destroy us is a circumstance where earth will be too near some other exploding supernova. If any living things survive, the struggle for survival will virtually start from point zero again.

SECTION V
THE DEATH OF STARS

The death of stars is important to our future on this planet. Only through violent explosions at the end of a star's life can the material of which so much of our planet is made be fed back into space where it then becomes available for the next generations of star and planet formation. The remains of stellar explosions, the supernovae, cast out beautiful nebulous material and their cores collapse to form neutron stars or black holes. Other stars end their lives less violently; these shed an outer shell of matter that moves out into space to form the beautiful planetary nebulae.

Although the existence of black holes is not firmly established, black holes are clearly going to give scientists much fun during the years to come; the understanding of the physics involved might well lead to some radical changes in what we now believe to be the correct description of the universe. At present there are many areas in astronomy (big-bang theory, quasars, black holes) in which conventional physics seems to fail, and seeking understanding of these strange phenomena may yet lead to a revolution in thought.

CHAPTER 16
STAR DEATH

Once upon another time and planet, some 7,000 years ago and 6,000 light-years away, another civilization watched in awe as a bright new star appeared in their sky. It grew in brilliance very rapidly until, within a day, it was more luminous than their sun. Panic spread about that planet and a few days later strange new diseases became rampant and millions died. Soon the planet was in utter chaos. Their science had not developed enough to allow them to even guess about what was happening. After a few months there were few left alive to watch the new star slowly fade.

Six thousand years passed and the light from this bright new star reached the planet earth. On the morning of July 4, 1054 A.D., Chinese astronomers watched intently as a new star appeared in the constellation of Taurus, the Bull. They diligently recorded the occurrence of this star:

> In the first year of the Shih-huo period, in the fifth moon, on the day of Ch'uh-Ch'iu a guest star appeared about several inches southeast of T'ieng'K'uang. Over a year later it gradually became invisible.

Eventually 20th century astronomers would study the remains of the stellar explosion that had produced the guest star and they would learn how the remains of exploded stars were capable of emitting so much energy.

The remains of the guest star of 1054 are called the Crab nebula, which is located in the constellation of Taurus. The guest star seen by the Chinese was a supernova, the annihilation of a star in a cataclysmic explo-

118   THE DEATH OF STARS

The Crab nebula, remains of the star that was seen to explode in 1054 A.D. Located some 6,000 light-years from the sun, this supernova remnant is one of the best studied objects in deep space.

sion. This new star, which brought death and destruction to all life forms within 30 light-years of it, brought good news to the Chinese.

> I make my kow-tow. I observed the phenomenon of a guest star. Its color was slightly irridescent. Following an order from the Emperor, I respectfully made the prediction that the guest star does not disturb Aldebaran: I beg to store this prediction in the department of historiography.

Was this astronomer keeping a scientifically open mind on the consequences of the supernova of 1054 A.D.? A bad prediction would have had negative consequences for the stability of his head on his shoulders! Expediency was perhaps the better part of valour. In those days one followed orders given by the emperor. Although modern astronomers can trust the location of the guest star recorded by the Chinese astronomers, they place little weight on the color given!

The supernova in Taurus was one of four seen in the Milky Way galaxy by humans since 1000 A.D. One was observed in 1006 A.D. and was regarded as the cause of wars, death, and pestilence among the Muslims at that time. The others were seen in Europe in 1572 (as noted by Tycho Brahe) and in 1604 (noted by Johannes Kepler).

Supernovae now appear to be the most important astronomical phenomenon as far as our lives are concerned. They are the source of both the elements needed to build planets and the molecules in our bodies. They influence our daily lives and will probably be the end of all life on this planet. How is all this possible?

The stars that have become supernovae are thought to be the cosmic cauldrons in which all the heavy elements, which are the basis for our life forms, are cooked up. The evolution of an interstellar gas cloud into a star, and ultimately into a gigantic explosion as the star suddenly dies, reveals how the elements in the universe are built up from hydrogen gas. Let's follow this process briefly. Hydrogen gas is the most basic matter in the universe and exists in giant clouds throughout space. Sometimes these clouds coalesce to form groups of stars. When these stars heat up enough to trigger nuclear reactions, their hydrogen starts to burn, forming some of the heavier elements such as helium. As the star ages and uses up its hydrogen, its interior sometimes shrinks and heats up more so that the temperature rises even higher; subsequently, helium burns to form other elements such as carbon. It might take billions of years to exhaust the star's hydrogen supply in this element-forming process.

Someday, when the interior gets very hot (over 100 million degrees Kelvin) new processes occur and suddenly the star will burn up carbon and produce heavier elements, e.g., iron. A point is finally reached when, in a few minutes, the core of the star is suddenly converted to iron and releases enormous amounts of energy, shattering the star. Simultaneously, the iron is converted into other, even heavier elements. But this process requires external energy for its maintenance. As a result, the core of the exploding star rapidly cools and collapses.

The resultant shattering of the star spews the heavier elements out into space in a shell of gas. This shell will ultimately expand so far that it will blend with surrounding matter and no longer be recognized as the remains of a supernova. Thousands of old and diffused shells exist in the galaxy, which means that space is pervaded by matter from old supernovae. After billions of years, interstellar space is, therefore, uniformly filled with heavy elements essential to the formation of planets (which consist of much iron in their cores) and essential for the formation of life based on molecules containing these heavy elements.

For life to barely begin in the galaxy, elements would have had to be formed in several generations of stars, many of which must have exploded violently in the process. Only when there is sufficient stellar debris that gets further processed into new stars can earth-like planets be formed.

Every atom in our bodies has been through a star at least once, and every heavy atom in our bodies has probably been formed in a supernovae explosion. We are, indeed, children of the stars.

A nearby supernova might, at some time in the future, wipe out life on this planet, probably well before our sun dies of old age.

Before (June 6, 1950) and after (February 7, 1951) photographs of a supernova explosion in the galaxy NGC 5457 in Ursa Major. The arrow indicates the position of the exploding star that shines as brightly as the whole section of the galaxy.

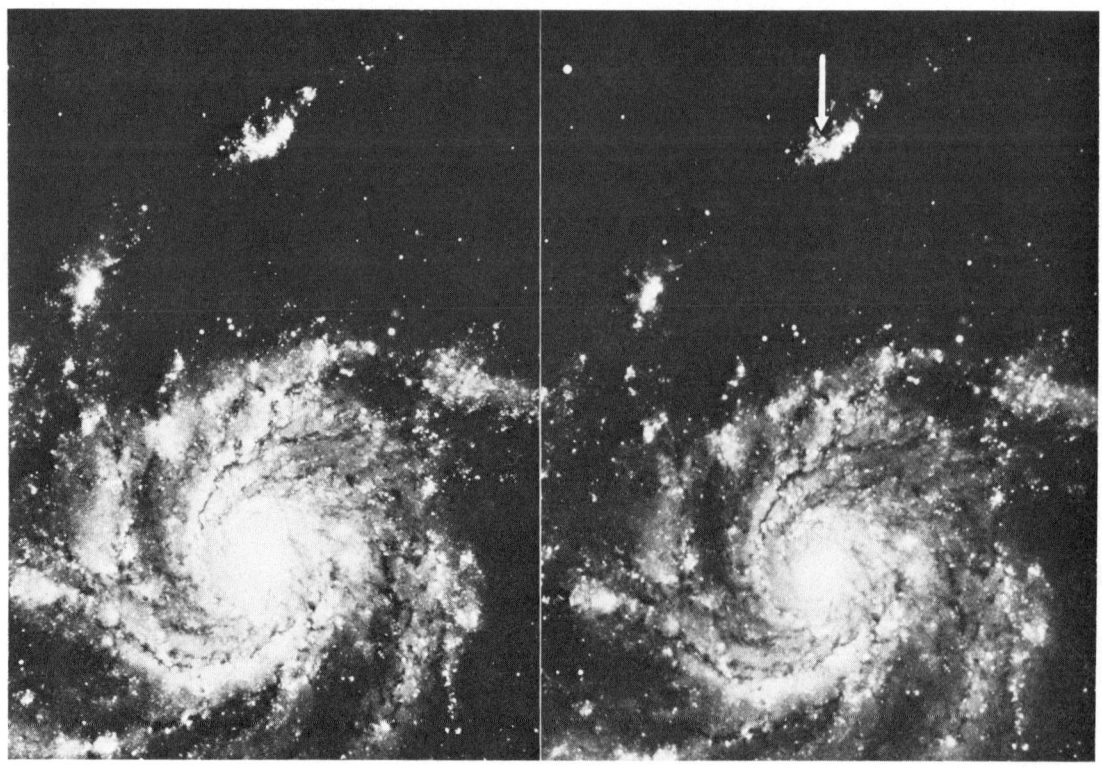

STAR DEATH 121

The filamentary remains of a star that exploded tens of thousands of years ago. Known as IC 443, this supernova remnant is a strong emitter of radio signals.

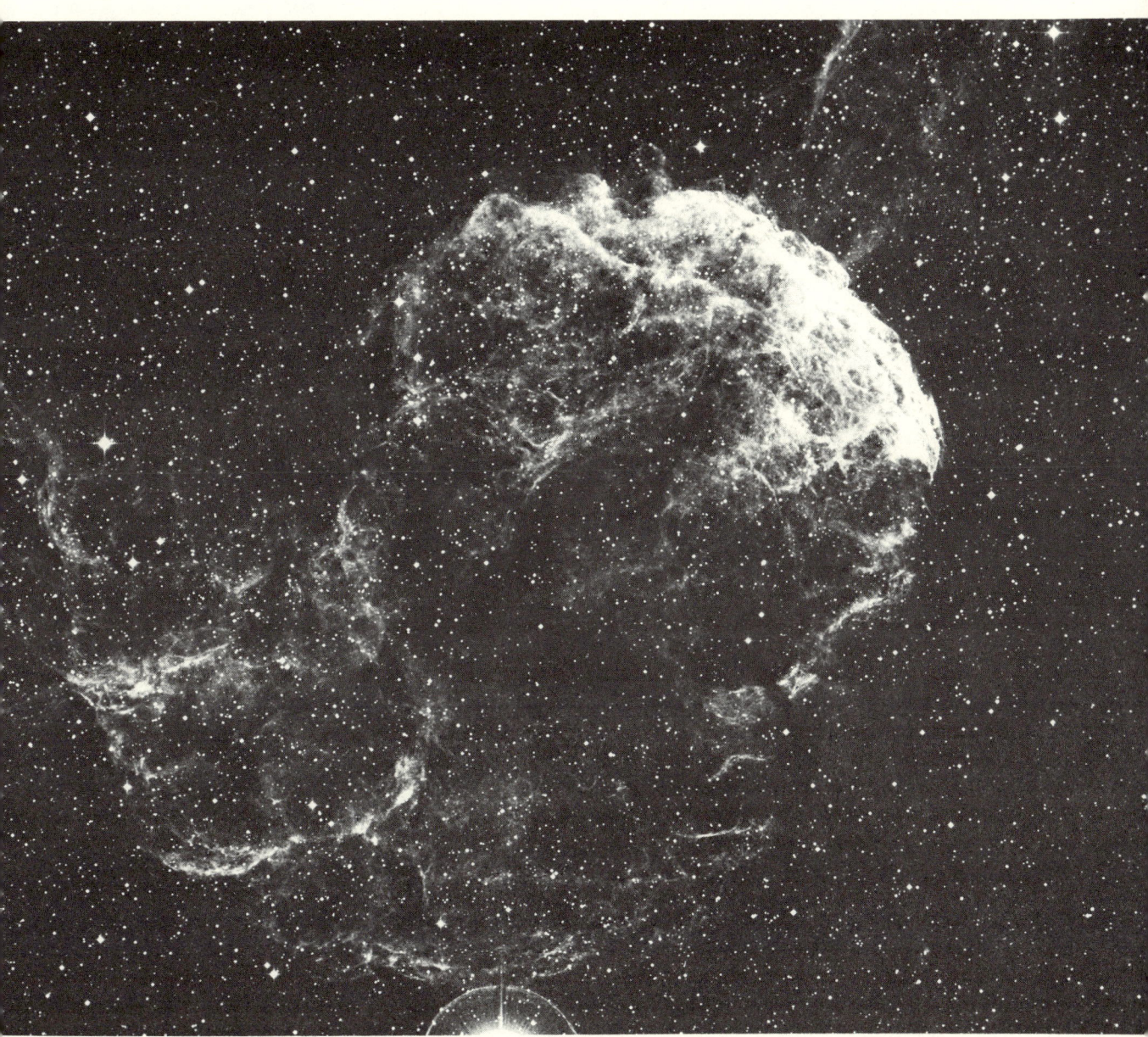

It is possible that nearby supernovae were the cause for the sudden extinction of various species of life on this planet in the past. We may even be in a race between humans and a nearby supernova to see which wipes out life on this planet first! The primary reason for this race is ozone, a molecule existing in the highest reaches of earth's atmosphere.

Ozone, a molecule built up of three oxygen atoms, protects us from the sun's ultraviolet radiation. In large doses, this radiation is fatal to humans and other life forms. Ozone also absorbs heat from the sun, controlling the temperature distribution in the high atmosphere and determining global weather patterns. A reduction in the naturally occurring ozone level will not only affect the weather, but will also allow more ultraviolet light to come through. This increases enormously the incidence of skin cancer. Some believe that a 10 percent reduction of ozone would probably produce at least 80,000 additional cases of skin cancer each year. Furthermore, the mutation of viruses in the earth's atmosphere depends heavily on doses of ultraviolet light that permeate the ozone layer. Remove or reduce that layer and mutations will speed up with unknown consequences.

Humans are in the process of wiping out this protective ozone layer because of the pollution we create on the surface of the planet. Pollutants, such as aerosol sprays, drift into the high reaches of the atmosphere and destroy the ozone. If a supernova occurred near us, say within 30 light-years, it, too, would severely affect, and possibly destroy the ozone layer, although it would take many years for this to happen.

Another way humans destroy the ozone is by injecting nitrogen oxides into the upper atmosphere; these compounds combine with the ozone and destroy it. Nuclear explosions and exhaust fumes from high-flying supersonic transport planes are sources of nitrogen oxides. Both sources are strongly opposed by environmentalists and humanists, but will they ultimately win the battle to preserve life on this planet?

There is also a natural source of nitrogen oxides in the atmosphere. Cosmic rays — the high energy particles that continually bombard us from outer space — are produced in supernova explosions and in the sun. At present there is a delicate balance between nitrogen oxides produced by cosmic rays and the ozone content, but if a supernova occurred within 30 light-years of the earth, this nitrogen oxide production would increase enormously (although delayed by perhaps 100 years after the explosion) and the ozone layer would quickly disappear. It would be fatal to stand in the sunlight because the ultraviolet dose would be lethal. Life would be wiped out on any planet within 30 light-years of a supernova; even if one occurred within 50 light-years, the additional doses of cosmic rays would almost certainly produce a health hazard, because of the depletion in the protective layer of ozone.

We mentioned the constant presence of cosmic rays reaching earth from outer space. If these particles are produced in supernovae explosions why do they reach us when no supernova has actually recently

The network of nebulosity associated with the pulsar and supernova remnant in Vela, a southern hemisphere constellation.

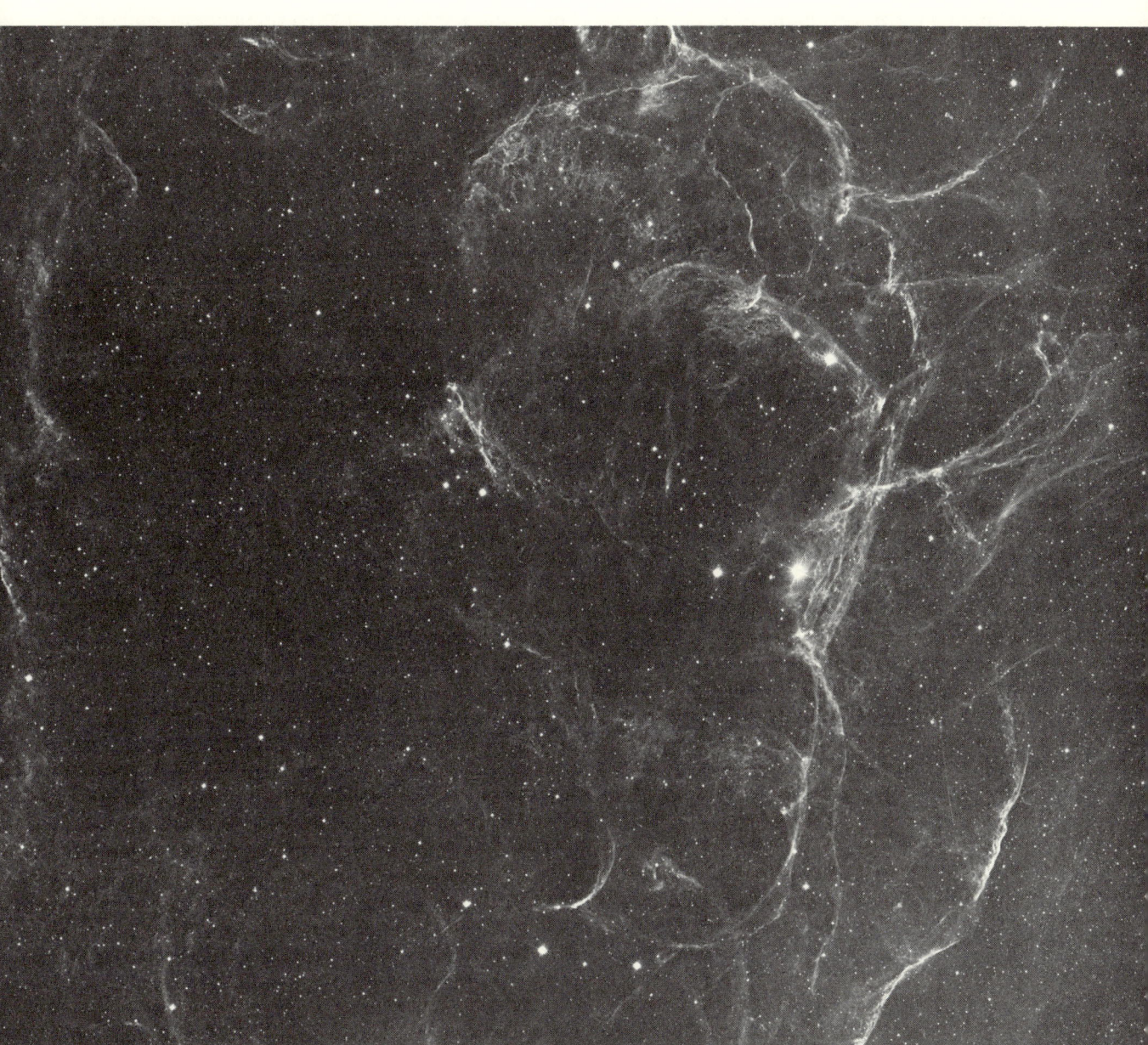

exploded near us? The answer is found by examining how cosmic rays are produced and how they travel through space.

In the initial stellar explosion two things happen. The first is the ejection of a shell of matter into space, a shell that contains many of the heavy elements formed just before or during the explosion. Second, the matter that was not thrown out as part of the shell rapidly cools and collapses in upon itself with tremendous force. As a result, the basic constituents of matter known as protons and electrons are forced together with such energy that they combine, forming neutrons. The remains of the star then become a very small ball of matter consisting almost entirely of neutrons. Stars of this composition that emit radio signals in pulses of incredible regularity are called pulsars. The neutron star (which may also be a pulsar) is expected to spin very fast as a result of its collapse to a small sphere some 10 km across (it was originally larger than the sun). This spinning neutron star becomes a constant source of energy for the nebular shell of matter that is now expanding out into space at thousands of kilometers per second. The energy of rotation can be fed into the surrounding nebula by one of several processes, each of which feeds the energy into actual particles of the nebula. These particles are the nuclei of the heavy elements, protons (the nucleus of the hydrogen atoms), and electrons. The energy available to these particles accelerates them until they travel at nearly the speed of light. When they encounter magnetic fields in the nebula, the particles spiral about these fields and, as a result, radiate light and radio signals. This process makes the nebula visible to us, even at great distances.

Some of these fast traveling particles, now called cosmic rays, leak out into surrounding space where they spiral about interstellar magnetic field lines as they travel between the stars. This means that the cosmic rays will ultimately be spread throughout the galaxy. Since at least one supernova is thought to explode in the galaxy every 50 to 100 years, and the injection of cosmic rays from the supernova remnants into interstellar space continues for possibly millions of years, we find that the galaxy is always full of cosmic rays.

These particles constantly bombard the earth and even now are passing right through our bodies, just as X rays can. In fact, if we were to receive a very heavy dose of cosmic rays directly, it would be fatal, just like an overdose of X rays. However, to receive such a dose from an exploding star, we would have to be very close to it, probably within its planetary system. Since the sun is not likely to become a supernova, we need not concern ourselves about that. Instead, the much more subtle effects produced by the reduction in the ozone layer due to cosmic rays from supernovae within about 30 light-years is what poses the potential danger.

A supernova should occur within 30 light-years of the earth once every 100 million years, and such an occurrence would severely affect the evolution of life on this planet. Evolution would suffer a severe setback, or, if the supernova was close enough, life might be destroyed entirely. The dinosaurs became extinct about 100 million years ago — could it be that a nearby supernova occurred and that some of the exposed life forms on this planet were killed by too much ultraviolet light?

We might conclude that to avoid fatal doses of ultraviolet light, we only need to stay out of the sun during the day. Perhaps we could simply cover streets with roofs and never venture into sunlight without protective clothing. But that would not be enough. The viruses that would be mutated by heavy doses of ultraviolet light can only be guessed at. We would probably have to live inside protective domes or caves, insulated from the outside world by sealed airlocks, with systems for extracting the dangerous viruses from the air. We would need space suits to wander outside, even at night.

And what of the climate? It appears that world weather patterns would be seriously altered. If the ozone level were to decrease drastically, the heat balance in the high atmosphere would change and the average temperature over the planet would rise. This would lead to a melting of the polar ice caps, causing flooding throughout the world.

Can we predict whether a supernova is likely to occur within a dangerous distance in the foreseeable future? Perhaps we can, if our present knowledge of the origin of supernovae is correct. If the supernova phase is entered only by relatively massive stars in their death throes, then we know that at present there is no obviously suitable star within 30 light-years. On the other hand, there is no assurance that our ideas about how stars become supernovae are entirely correct. It has been only ten or twenty years since astronomers have even pretended to understand the phenomenon.

Even as humans struggle to prevent their fellow humans from polluting the earth's atmosphere in ways that will irreversibly alter the environment and lead to countless deaths, a nearby stellar explosion could put an end to it in a few days. In the meantime, astronomers continue to study the universe as best they can and perhaps one day will learn exactly what a supernova is and what we should do when one explodes nearby. Inevitably some star will explode within lethal range of earth someday, and if we have not destroyed ourselves by then, our descendants will be able to watch in awe as the "guest star" grows to the brightness of the sun.

CHAPTER 17
WHY DOES THE CRAB NEBULA SHINE?

The Crab nebula was once a star. In the year 1054 that star destroyed itself in the violence of a supernova explosion. The eruption was recorded by Chinese astronomers and probably by American Indians in the southwest United States. It shone brightly for several months, so brightly that it was visible by day.

A supernova is obviously a very chaotic event, but we generally do not think about the way the light from such an explosion is generated. It seems obvious that it should produce a lot of light, since explosions generally do.

There are a few questions we should ask about various aspects of the supernova phenomenon. The first is: Why does it happen? Related to this we wonder which stars will explode. (This is sort of a personal concern since we are, after all, orbiting a star.) Another question is: What actually happens when the explosion occurs? That is, what generates the light, radio, and X-ray signals? Finally, why do we continue to see the supernova remnants after the explosion? In the case of the Crab nebula, it has been some 900 years since the blast. So why is it still shining? This question applies equally well to the remains of other supernovae.

We will try to give some answers to that last question here. (The first two questions are answered in chapters 15 and 16.) Since the Crab nebula is one of the most thoroughly studied objects in the universe outside the solar system, we will concentrate on it as our best example.

As long ago as 1945 Walter Baade discovered that photographs of the Crab nebula showed variations from year to year, and even from month to month. The changes were traced to the core of the nebula where several wisps of material seemed to be involved. This was the first indication that the Crab nebula was still "alive." Baade and Rudolf Minkowski suggested that one of the two stars lying near these wisps was possibly the collapsed remains of the star that originally exploded. We now know that they were right. The lower of the two central stars in the Crab nebula is a pulsar.

Pulsars are thought to be the highly condensed, collapsed cores of stars that have erupted in supernovae explosions. A colossal magnetic field riddled with intensely high-powered energy is wrapped around the 15 kilometer wide pulsar. The pulsar is, in effect, the power generator for the nebula. The way it injects energy into the nebula is quite fascinating.

The photo on page 128 of the central zone of the Crab shows "Wisp 1," a feature that actually moves backward and forward changing its apparent position by about two seconds of arc every couple of years. Since we know the distance to the Crab nebula (about 6,000 light-years), this can be interpreted as "something" moving with speeds as large as 30 percent of the speed of light. The wisp has been measured to move one-twentieth of a light-year.

In 1969 Jeffrey Scargle at Lick Observatory collected photographs that show the way this wisp changes its shape and position with time. He later found that these changes are directly related to activity in the pulsar itself.

The pulsar in the Crab nebula was discovered in 1968 as a result of the radio signals emitted by it. The pulses are regular, repeating once every 33 milli-seconds (that is, 30 times per second). After this discovery, the central stars were photographed with a camera system whose shutter was synchronized to the radio pulsar period. The star we identified earlier as the pulsar was observed actually flashing on and off at the same rate as the radio pulsar.

Even while pulsars are emitting very constant bursts of radiation, the repetition rate of the pulses is gradually slowing down from year to year. All pulsars are running down, much as a watch will do if it is not wound.

Then something unusual was discovered. The Crab pulsar suddenly sped up, and then slowed down to its previous rate. This was called a "spin-up" and it was thought that a sort of "starquake" had taken place on the surface of the pulsar. Perhaps the neutron star surface had realigned itself in some way, much in the way an earthquake will reshape the surface of earth. In the neutron star the quake might have caused the star to get a little smaller, which would cause it to speed up slightly.

After several spin-ups Scargle studied his photographs of the Crab nebula and its wisps, and found that changes in the wisps occurred about sixty to eighty days after each spin-up. He concluded that whatever had caused the spin-up also caused something to move out from the pulsar (that is, the neutron star), and when that something reached Wisp 1 it pushed the wisp outward. Later Wisp 1 would move back to its original position. Sometimes the more distant wisps would also show changes in structure.

Careful thought indicated that no starquake could generate a phenomenon that would travel the distance through space out to the wisp. Instead, it appeared that something else had caused both the changes in the wisps and the spin-up.

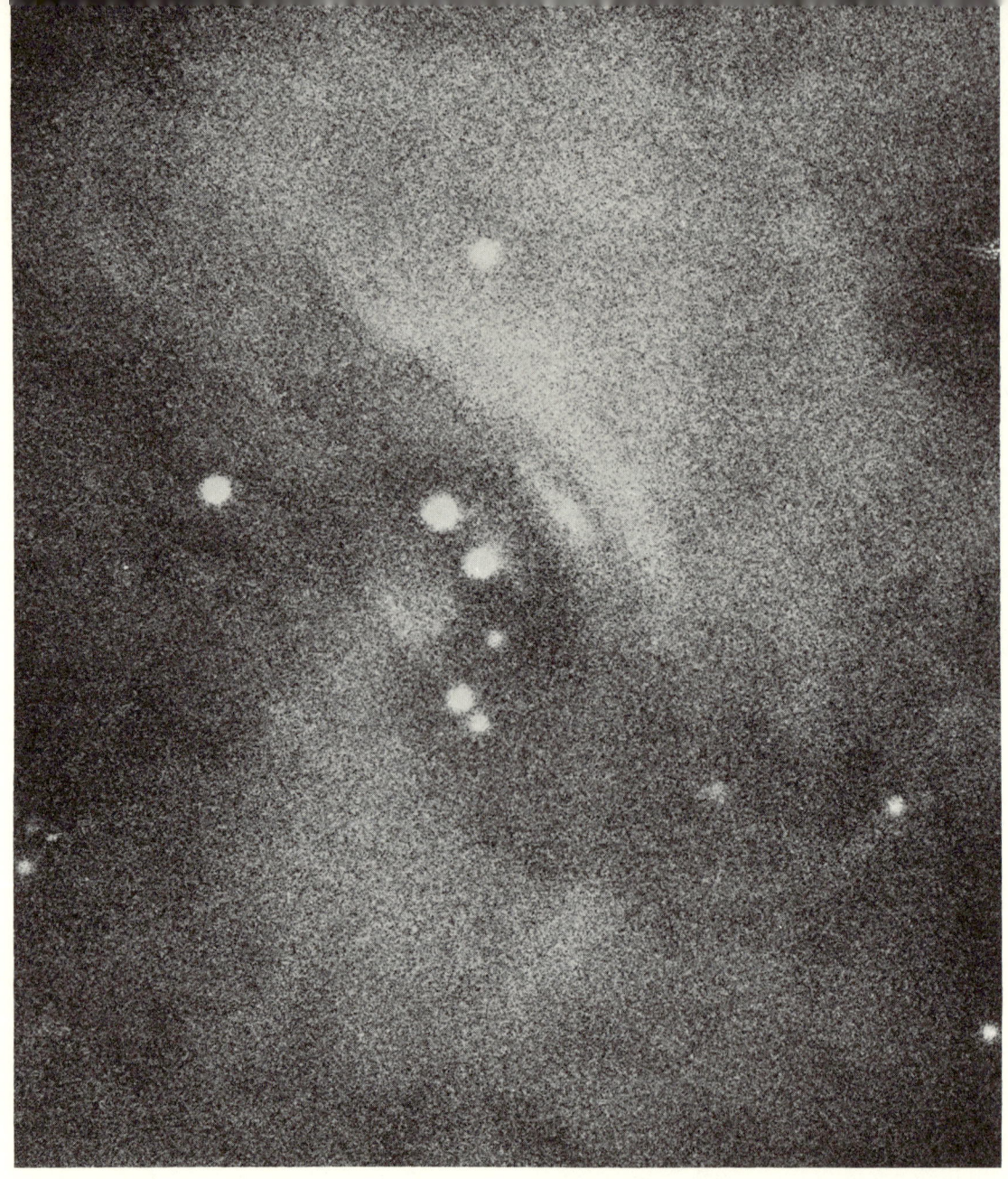

The central part of the Crab nebula showing the two bright stars near its center. The lower one is truly within the nebula and is known to be a pulsar. The bright wisps of matter near this pulsar change from year to year showing that energy is being fed into the nebula by the pulsar. This energy occasionally pushes the wisps around and causes them to brighten.

We know a pulsar is spinning very fast and we also know it has a strong magnetic field that gets continuously twisted as the pulsar winds it up. This will cause it to become unstable, which in turn will occasionally produce the equivalent of an explosion in the upper parts of the pulsar magnetosphere (the most intense part of the magnetic shells around the pulsar).

Such an explosion or rearrangement of magnetic fields would be similar to an explosive flare on the sun. As a result of this explosion, a wave of magnetic fields and particles moves out from the pulsar. Meanwhile, Wisp 1 is sitting out there. The blast wave from the pulsar hits it, and suddenly the wisp gets violently pushed around. At the same time, the abrupt change in the magnetic conditions around the pulsar cause it to momentarily speed up and then it continues to slow down as before. This slowing down continues for millions of years. Spin-ups have also been observed in other pulsars so we can assume the general slowing down is normal performance for these bizarre objects.

We are partly on the way to understanding where the energy for the nebula comes from. When the compression wave or blast wave from the pulsar hits Wisp 1 it can cause great changes in the magnetic field strength and structure in the wisp. Consequently, conditions in the wisp become suitable for accelerating particles such as electrons, and these particles then spread out into other parts of the Crab nebula itself. There they spiral around the magnetic fields in the filaments that are so clearly seen in the nebula. They then radiate radio and light energy, which we pick up here on earth. Some of these accelerated particles also leak out into space and form part of the cosmic ray background in the galaxy.

To summarize: Reorganization of the magnetic fields near the pulsar (a few kilometers from it) is probably what causes magnetic energy to propagate outward. It is then converted to energy in the motion of particles in the wisp. In turn, it is converted to electromagnetic (light, radio waves, and so forth) radiation that we can see here on earth. In 1976, an effect just opposite to spin-up was observed in one pulsar. This threw the theorists into a state of consternation; at this writing, no explanation for the neutron star phenomenon has accounted for all the data.

To get back into the center of the nebula, though, this is an intermittent process and not the only process that causes energy to get out into the nebula. The pulsar itself is constantly spinning — some 30 times per second. It is thought that the magnetic axis of the pulsar is not aligned with the rotation axis, much as earth's magnetic north pole is not at the geographical north pole. The spinning pulsar's magnetic field becomes a generator of 30 cycles per second radiation. Of course we know the pulsar sends out pulses at this frequency, but we are now talking about something different. The pulses probably come from regions above the magnetic poles and every time we look down certain magnetic field lines, we pick up a pulse. The 30 cycles per second radiation we must now consider is an actual electromagnetic wave, just as a radio or light wave is an electromagnetic wave of a certain frequency.

This 30 cycle wave finds it difficult to go far, and it gets absorbed by the matter in space around it (which is, of course, the matter in the Crab nebula itself). This means that the pulsar's energy of rotation is converted into a wave that gets absorbed by the nebula, where it is converted into energy. This is energy of motions of electrons and protons in the nebula, which means that they contain energy available for emission of light and radio signals.

The actual surface structure of pulsars is not accurately understood, but astronomers believe that the spinning pulsar magnetic fields cause currents to flow on the surface. The 30 cycle per second radiation is produced by these currents. At the same time, some of the electrons that make up these currents will blow out into space in the form of a strong wind. The sun also generates a constant wind of matter into space, but by a different process. The winds from the pulsar are available for feeding even more energy into the surrounding nebula. The first process mentioned above — that of the sudden instability — can be thought of as a sudden increase in the basic pulsar wind that flows out into space until it loses all its energy. It is in that region of space that the wisps are thought to exist — the region where the wind exhausts itself and matter piles up against the surrounding nebula.

The existence of these three processes is inferred by careful study of mountains of data and theory application. Now that it's done, we can understand, for the first time, why the Crab nebula is still shining brightly out there, some 6,000 light-years away, 900 years after the explosion.

CHAPTER 18
THE DUMBBELL, THE OWL, AND THE ESKIMO

This is not the title for some mystery thriller or a new fairy tale; these are names of planetary nebulae. There are a few others like the Ring and the Helix; but besides these, most planetary nebulae have the usual sterile astronomical names ranging from NGC 7027 through designations such as K 3–50, M 1–11, and BD 30° 3639, which hardly invoke any images in the mind's eye. But as the more descriptive names suggest, planetary nebulae are among the most dramatic and beautiful of all sky objects.

These celestial oddities were originally called planetary nebulae because their outlines appear disklike in most photographs and to the eye seemed to resemble the disks of planets. However the name is a complete misnomer since they are in no way related to planets. Instead, they are large, diffuse, spherical clouds of gas that have been puffed off by hot, dying stars, and heated by those stars so that they radiate their own light and radio waves.

There is a definite difference between planetary nebulae and the apparently somewhat similar emission nebulae, such as the Orion nebula. Surrounding very hot, young stars, emission nebulae are clouds of hot gas left over after stars formed within them. Planetary nebulae, however, are much smaller clouds of gas ejected by stars as they approach the end of their lives. The sun is expected to have a planetary nebula shell around it in about 6 billion years, after it has passed through the red giant stage (see chapter 15). The only similarity between the two types of nebulae is the fact that gas is heated by ultraviolet light from the stars themselves; the hot gas then radiates visible light, infrared radiation, and radio waves as a result of collisions between particles in the heated gas.

132   THE DEATH OF STARS

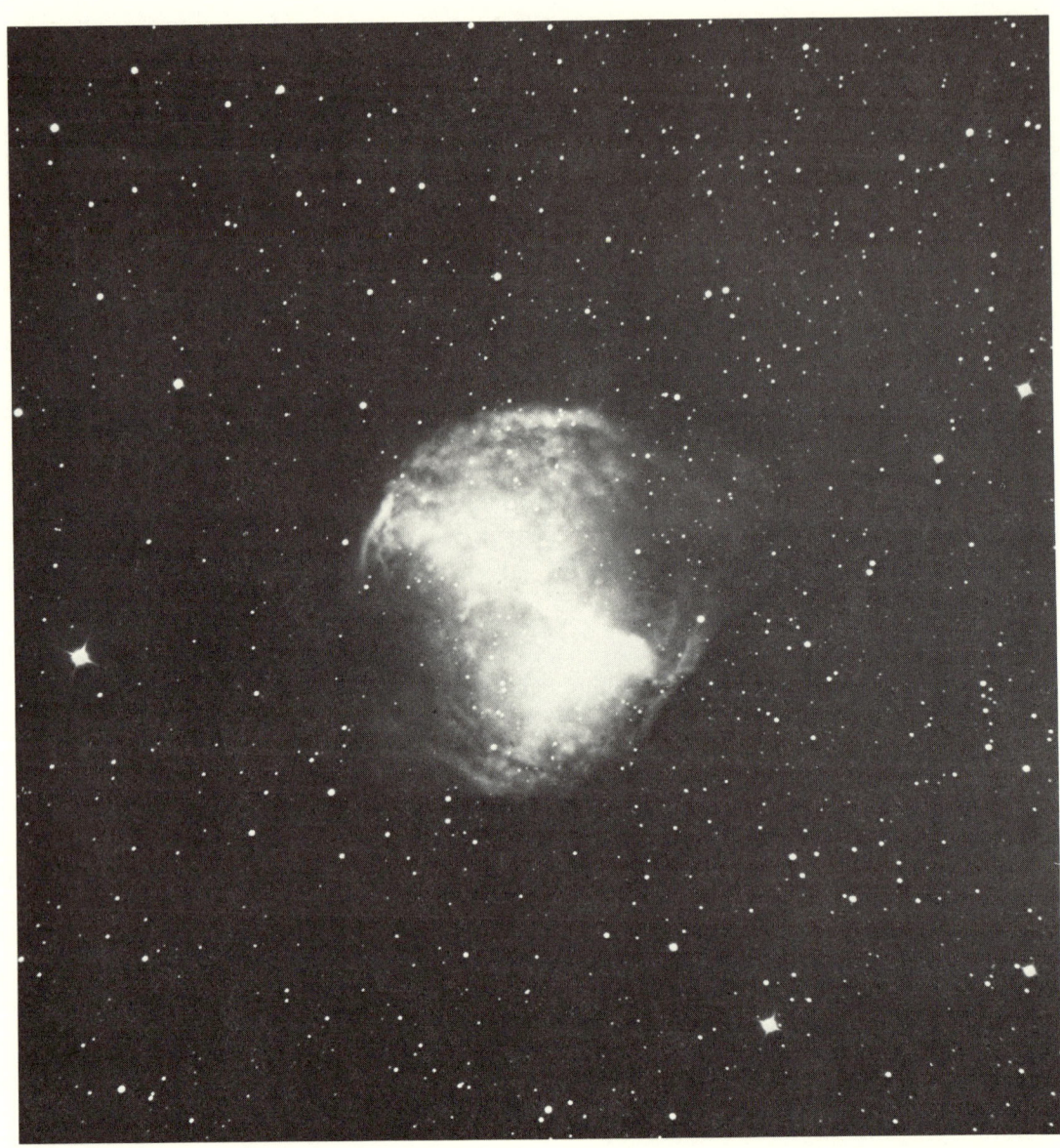

The intricate detail of the Dumbbell nebula can only be seen photographically — it looks more like its name through a small telescope. One of the most prominent planetary nebulae in the sky, this shell was ejected by a star visible as a small, bright white object in the middle of the nebula.

The central star in a planetary nebula may be as hot as 100,000 degrees Kelvin and normally is a small white dwarf. Despite their enormous temperatures they emit only as much total energy as the sun because the stars are not much larger than earth. The surrounding gas shell that forms the nebula is heated to about 20,000 degrees Kelvin and the gas is fully ionized (that is, electrons are stripped away from the atomic nucleus).

The size of the planetary nebula can range from a small fraction of a light-year in diameter to about one light-year across. Their angular sizes on the sky range from a few seconds of arc to ½ degree across, depending on age (young ones are the smallest) and distance. A typical planetary nebula is about 20,000 years old.

In trying to understand both the processes that occur inside these nebulae and their odd range of shapes, astronomers observe them at many different wavelengths. Radio and optical observations are combined with data collected at the specific wavelengths of atoms such as oxygen and nitrogen. For example, the planetary nebula known as NGC 7027 has an infrared picture exactly the same as the radio picture, but the optical picture looks different. The reason appears to be that the optical picture is partly affected by obscuration from interstellar dust between us and the nebula, but the radio and infrared radiation pass unimpeded through the dust. However, although the radio signals come from the hot ionized gas, the infrared radiation is known to originate from hot dust particles. Since the two pictures of the nebula look alike, the hot dust and hot gas must be mixed in the same volumes of space. We might be watching dust particles being injected into space from a star — particles that will ultimately blend with the dust already in the rest of space.

Photographs of planetary nebulae taken in the light radiated by various atoms in the nebulae often show that these objects look different at different wavelengths. This is because the physical conditions required to make one atom (or ion) radiate differ from those that cause another atom (or ion) to radiate. Study of this phenomenon reveals that small blobs of high density and relatively cold matter are situated inside most planetary nebulae. These blobs are so dense that the ultraviolet radiation that usually heats the gas cannot penetrate to the inside of these small condensations. In fact, the globules can cast a shadow within the nebulae so that matter beyond these small blobs, on the side away from the star, might be cooler than unshaded matter. Thus, the nebulae can take on all sorts of weird shapes, such as is seen in the Eskimo nebula.

Why should such small blobs of concentrations form? After all, we think of planetary nebulae — such as the Ring nebula (shown on page 109) — as being nicely symmetrical objects. Indeed, at first sight most planetary nebulae are symmetrical and often round, although somewhat elongated. The elongation appears to have something to do with surrounding magnetic fields that influence the way the ejected shell of matter expands. Let's take a close look at what does happen to the expanding shell of matter as it is ejected.

In the beginning the star might find itself in a cloud of interstellar matter — the common neutral hydrogen clouds that exist throughout interstellar space, even around

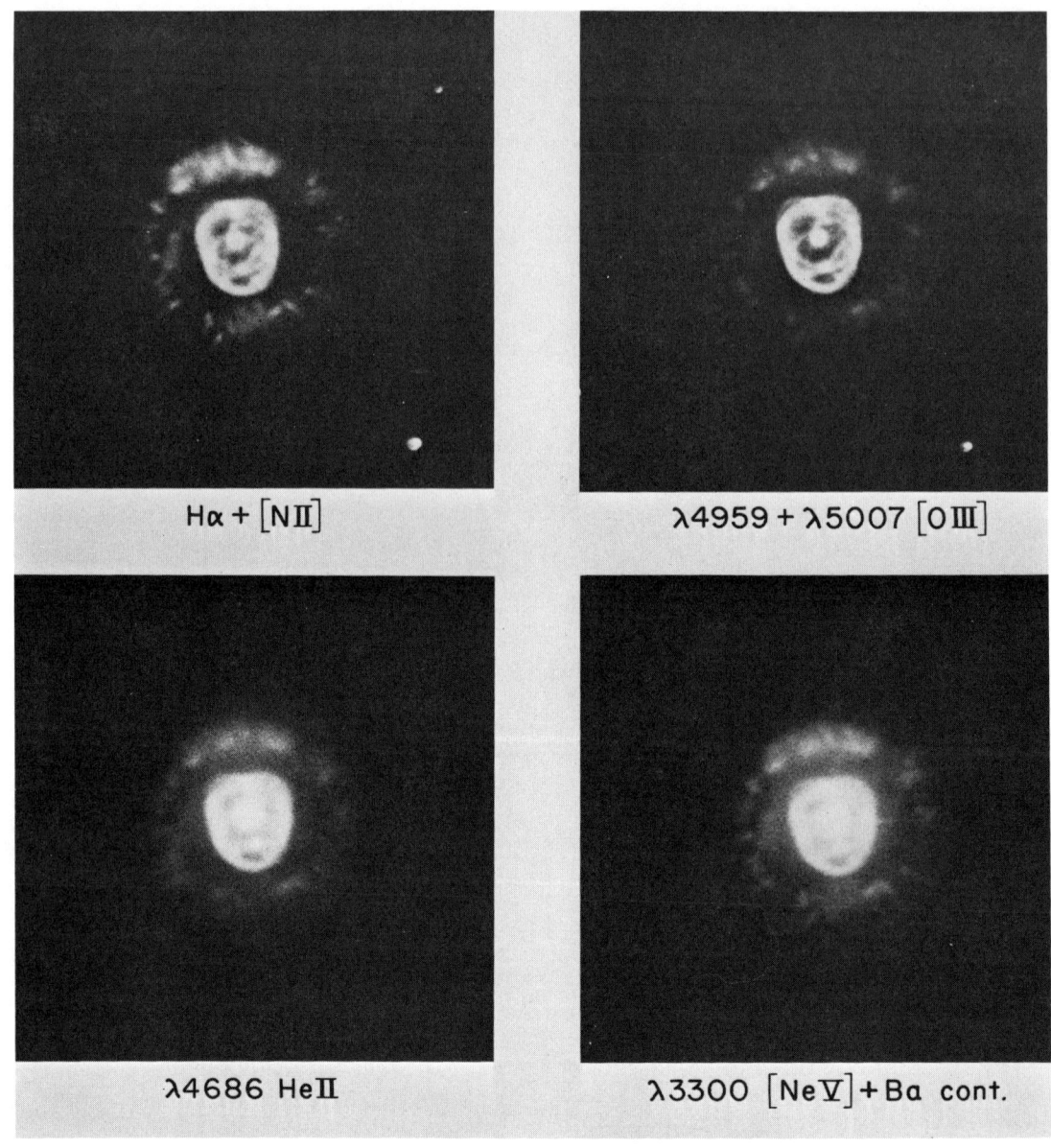

The planetary nebula NGC 2392, sometimes known as the Eskimo nebula, photographed in four different wavelength regions. Different parts of the nebular material are revealed, depending on their temperatures.

The planetary nebula NGC 3587, also known as the Owl nebula, in the constellation of Ursa Major.

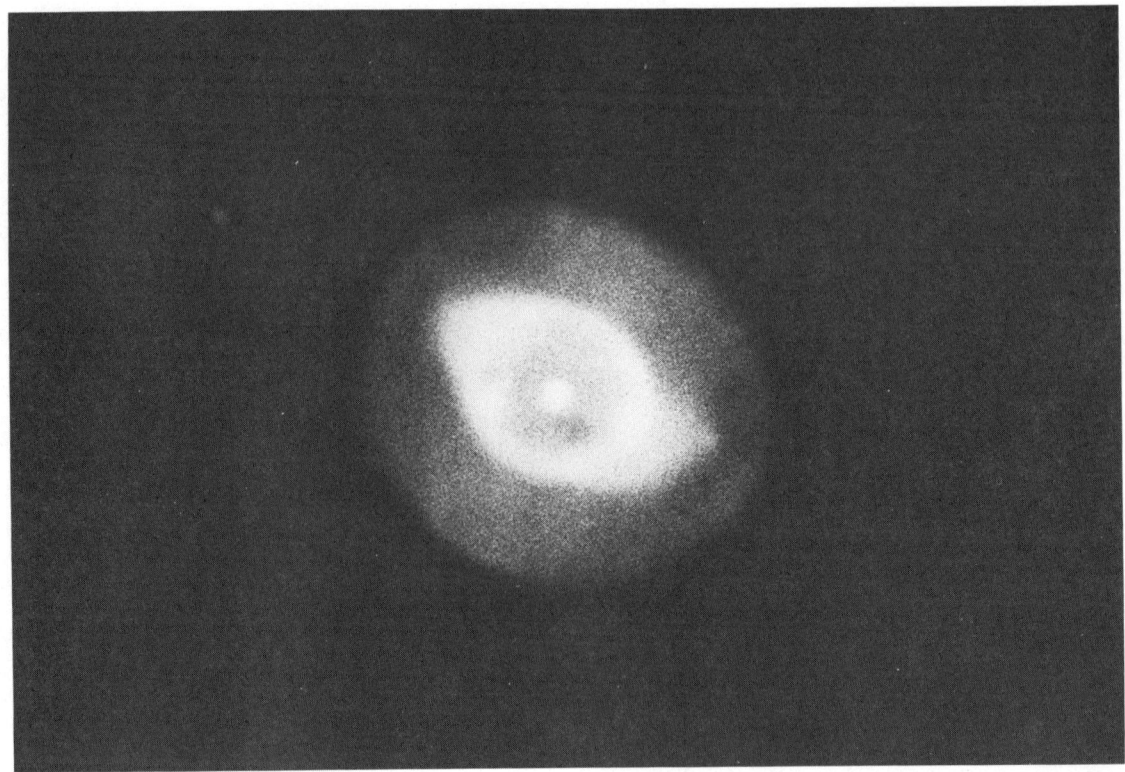

The planetary nebula NGC 3242 in the constellation of Hydra. The central star that ejected the shell of matter is located precisely at the nebula's center.

the sun. When the star ejects its shell of gas, the energy in this hot gas is great enough to force the surrounding cold interstellar gas (about −100 degrees Kelvin as compared with 10,000 degrees Kelvin) outward like a piston. As the expanding shell gets larger, the ultraviolet radiation that heats and ionizes the gas has to travel farther and farther from the star. This radiation will ultimately get used up on its way out to the edge of the nebula just to be able to maintain the high temperature in the nebula. Around this time the pressure of certain areas in the surrounding interstellar gas will start to balance the pressure within the planetary nebula; the shell then becomes unstable. It is something like the effect you get when you push in the sides of a balloon as it is being inflated. Parts of it are prevented from expanding while other areas bubble out. So it is with the planetary nebula shell. As soon as one spot is pushed in a bit more than any other spot around the shell, the whole thing gets distorted rather quickly.

Such a distortion quickly spreads and is known as an instability. As a consequence, a lot of spikes of hot matter will push out into the surrounding gas cloud while the zones of cold matter knife inward into the

The planetary nebula NGC 7293 in Aquarius showing remarkable small-scale structure in the shell ejected from the central star.

hot nebular shell. As these hot and cold spikes of matter rush past each other they further distort and start to form small condensations of dense matter, globular in shape. Calculations have shown that as many as 100,000 of such small globules may form. About half of the matter in the nebula may end up in such condensations. These condensations are so dense that they are shielded from the ultraviolet radiation of the star. They will therefore shade other parts of the nebula. As the nebula ages and ultimately ceases to exist after 100,000 years, hundreds of thousands of small globules will be fed into the interstellar medium.

Clearly, the shape of any given planetary nebula will differ depending on its stage of evolution and the distribution of globules within it. Then, since different radiations originate in the shadowed regions as compared to the illuminated parts of the nebula, the planetary nebula will look different when examined in the light from different atoms and ions.

One of the strangest types of planetary nebulae are those that have an apparent helical shape; that is, they look like parts of a helical spring seen on end. The Helix nebula, NGC 7293, is one such object while NGC 6543 is another. Parts of NGC 6543 are moving away from the central star at about 10,000 kilometers per hour on both the far and near sides of the star. This is faster than most planetary shells expand and must be due to youthful vigor. Various other calculations confirm this and give this nebula an age of only 1,000 years. In another 10,000 years it will look like a typical amorphous planetary nebula showing little structure.

Despite the large number of dramatic shapes exhibited by planetary nebulae, some are featureless. The Palomar Sky Atlas has dozens of planetary nebulae that look like blemishes on the photo plates! On the other hand, objects such as the Ring nebula are dramatically beautiful and can be seen quite easily with small telescopes.

Where are most of the planetary nebulae in our galaxy? They are where most of the stars are — in the densest regions of the Milky Way. There may be as many as 50,000 planetary nebulae in the galaxy, which is only a very small fraction of the 150 billion stars in the entire Milky Way galaxy.

SECTION VI

GALAXIES

CHAPTER 19
ISLANDS IN THE SKY

Imagine living on an island some 10 kilometers across. The nearest islands are only a few kilometers in size located about 15 kilometers away. The next islands about the same size as your own are located about 200 kilometers away. In between there are a few much smaller ones whose distances are about 30 or 40 kilometers. Beyond the nearest twin island, you have to travel another 400 or 500 kilometers in any direction before you reach any other islands. This roughly resembles the location of our Milky Way galaxy in space.

The Milky Way galaxy is an island floating in space. Shaped like a flat disk 100,000 light-years across, it contains several hundred billion stars. Near it are several smaller, dwarf galaxies, and in the distance we see the galaxy that is often called the Andromeda nebula or M 31. The prefix refers to the catalog of Charles Messier. In 1781, Messier, an ardent comet hunter, prepared a list of things to avoid when you looked at the sky in search of comets. He listed 103 objects, and the galaxy in Andromeda is number 31 on his list. Many of these "things-to-be-avoided" are galaxies, and the remainder are nebulae such as supernovae remnants or emission nebulae like the Orion nebula (M 42) in our galaxy.

M 31, somewhat larger than the Milky Way, is located about 2 million light-years from us. It, too, is attended by a swarm of galaxies, including the relatively large one known as M 33. In all, about twenty nearby galaxies are floating around together, forming what is known as the Local Group. Our nearest neighbors in the Local Group are the Magellanic clouds (160,000 light-years from us), which can be clearly seen in the southern hemisphere skies as hazy patches of light near the south celestial pole.

142 GALAXIES

The galaxies we are familiar with today have only recently been shown to be outside the Milky Way itself. Before 1924 most astronomers thought that all the nebulae, as they were then called, were members of our own galaxy. However, the concept that there were island universes like the Milky Way, spread throughout space, had already been formulated by Immanuel Kant in 1755. The concept of island universes was appealing, but no one was able to prove its validity until the distances to some of the nearer galaxies were measured and were shown to be very far outside the Milky Way (chapter 13). At first, the name extragalactic nebulae was coined, referring to the nebular or cloudy nature they exhibit when viewed through a telescope. The Andromeda galaxy, M 31 (see p. 87) looks like a faint cloud when seen with the naked eye. Later the name galaxy was applied to these extragalactic nebulae. They were all found to be enormous conglomerations of stars. (The term nebula is still applied to clouds of gas and dust,

The striking spiral galaxy M 101 seen head-on. Spiral arms containing stars, gas, and dust are clearly evident in this galaxy. In the outer regions they obviously lose their ordered structure.

illuminated or heated by stars, which exist between stars).

The discovery of what is now known to be hundreds of billions of galaxies in the universe placed earth even farther into the background as far as its importance on a cosmic scale is concerned. Although Copernicus finally convinced us that the earth was but one planet orbiting the sun and we were not the absolute center of the universe, we continued to believe that at least we were placed in an important location in the Milky Way. The Milky Way galaxy was the fundamental group of stars we knew about. Now we know we are located some 30,000 light-years from the center of our galaxy. The revolution in thought that accompanied the removal of the earth from the center of the solar system had no parallel when the Milky Way was removed from its prime position in the universe. Carrying this further, some astronomers are considering the existence of multiple universes; the one we can see through our telescopes just happens to be the one in which we live. Beyond the range of present-day telescopes and beyond our time, there may well be other universes, although no one has yet figured out much about them, or how we can, if ever, see them.

When we read relatively old articles about astronomy written forty or fifty years ago, we are often struck by how little was known; we are also aware of how much we have since learned. With what perspective will someone fifty years from now read our astronomical journals and books? Will our knowledge of the universe have changed so much that most of these chapters will be totally out of date, possibly even wrong? Or is the flood of new knowledge we have gathered over the last fifteen years nearly over? Whatever the answer is, I feel that in the area of understanding galaxies we might well leave present ideas farther behind than in any other area of astronomy.

Unlike the area of planetary astronomy, in which we are now traveling to the planets, or the area of stellar astronomy, where an enormous amount is being learned about stars with the aid of great new telescopes and enormous computers, it seems that the field of galaxy research is still in its infancy. The origin of galaxies is not yet understood, nor are the processes that determine what happens to the general outline or structure of a galaxy as it evolves — if it does evolve.

Some obvious types of galaxies do appear in the sky. There are many of the class known as spiral galaxies that contain dust and gas, and both old and young stars. The gas, dust, and young stars are ordered in streamers of material called spiral arms. Sometimes these spiral arms can be followed around half of the galaxy; at other times only bits and pieces of spiral structure can be seen. In almost all cases, little feather-like projections can be seen sticking out of the spiral features. At present theoreticians are having an interesting time trying to explain why some spiral galaxies — perfect spirals that is — have the shapes they do. When we look at photographs of galaxies we find that few have the perfect spiral pattern that the theoreticians are trying to explain. But, of course, in any area of research we have to start somewhere.

Other galaxies show no detailed structure at all. They are simply large, uniform conglomerations of stars, often spherical or elliptical in outline. These galaxies contain

The elliptical galaxy NGC 4486, also known as M 87, in the constellation of Virgo. This galaxy emits strong radio signals, although optically it shows little structure at all. It consists entirely of old stars, with no associated dust or gas. A large number of globular clusters can be seen surrounding this galaxy. A peculiar jet emerges from the center of this galaxy, but in this exposure it is almost completely lost in the brightness of the galaxy's central regions.

almost exclusively old stars; no gas, dust, or young stars. As a class these are called elliptical galaxies. There is no obvious evolutionary connection between spiral and elliptical galaxies.

Other galaxies have structures of a very irregular nature; no spiral pattern is visible in them, nor are they smooth like ellipticals. These are called irregular galaxies. The Magellanic clouds, our galactic neighbors, are examples of this type.

The single most important glimpse of galaxies we can get is by looking at a collection of photographs (a catalog), Peculiar Galaxies, collected by Halton Arp of the Hale Observatories. After paging through his collection it is easy to lose faith in what is now taught about galaxies in standard astronomy books. Our view of the classification of galaxies may be heavily biased by a few early catalogs prepared, for example, by Hubble.

The spiral galaxy NGC 4565 — an example of a galaxy seen edge-on. The presence of dust clouds in this galaxy is evident.

He examined the nearest, largest, and some of the most ordered-looking galaxies around. Undoubtedly there are many ordered-looking galaxies in the universe, some of which show spiral patterns, but there are many pathological cases in the photo files. The question is which are normal, the odd-looking ones or the neatly patterned ones?

Hubble originally classified 50 percent of the galaxies as spirals. But if we examine only the Local Group (see Table 19.1) we find that of some twenty members, only three are spirals. The rest are ellipticals, mostly small ones, and irregular galaxies. It is quite possible that such a distribution is also true for other groups or clusters too far away to see many of their members.

The largest galaxy in the Local Group is M 31 (see page 87). It is a large spiral galaxy containing the equivalent of 300 billion times the sun's mass (our galaxy contains about 200 billion solar masses of material). M 31 is inclined at about 15 degrees to the line joining it and us. The presence of hydrogen gas was found from radio observations revealing that the hydrogen stretches into space well beyond the region where stars can be seen in M 31. The optical

TABLE 19.1 The Local Group of Galaxies

Galaxy	Type	Visual Magnitude	Distance (1,000 light-years)	Diameter (1,000 light-years)	Approximate Luminosity (million suns)	Approximate Mass (million suns)
Milky Way	Sb	—	—	100	7,600 ?	200,000
Large Magellanic Cloud	Irregular	0.1	160	30	2,000	6,000
Small Magellanic Cloud	Irregular	2.3	195	25	400	1,500 ?
Ursa Minor System	E4 (dwarf)	10 ?	230	3	0.25	0.25 ?
Sculptor System	E3 (dwarf)	8.0	260	7	4	3 ?
Draco System	E2 (dwarf)	11.5 ?	330	5	0.2	0.2 ?
Fornax System	E3 (dwarf)	8.4	600	10	20	20 ?
Leo II System	E0 (dwarf)	12.0	620	4	0.4	1 ?
Leo I System	E4 (dwarf)	12.0	1,300	7	2	2 ?
NGC 6822	Irregular	8.9	1,500	9	150	1,500 ?
NGC 147	E6	9.7	1,950	10	70	350 ?
NGC 185	E2	9.4	1,950	8	90	450 ?
NGC 205	E5	8.2	2,250	16	275	1,500 ?
NGC 221 (M 32)	E3	8.2	2,250	8	275	1,500 ?
Andromeda Galaxy (M 31)	Sb	3.5	2,250	130	21,000	300,000
Andromeda I	E0 (dwarf)	14 ?	2,250 ?	1.6 ?	2 ?	2 ?
Andromeda II	E0 (dwarf)	14 ?	2,250 ?	2.3 ?	2 ?	2 ?
Andromeda III	E3 (dwarf)	15 ?	2,250 ?	2.9 ?	2 ?	2 ?
NGC 598 (M 33)	Sc	5.8	2,250	50	3,000	8,000
IC 1613	Irregular	9.6	2,400	18	60	400 ?

The table lists those galaxies that are generally agreed to belong to the Local Group. Particularly uncertain quantities are followed by question marks.

diameter of M 31 is about 130,000 light-years, somewhat larger than the accepted value for our galaxy (about 100,000 light-years across).

M 31 is accompanied through space by two small elliptical galaxies known as NGC 221 and NGC 205 (NGC stands for New General Catalog). These are but two of seven satellites of M 31, just as the two Magellanic clouds are thought to be satellites of our galaxy. Unlike NGC 221 and NGC 205, the Magellanic clouds are irregular galaxies containing much hydrogen gas and young stars. The two Magellanic clouds are separated by about 30 degrees on the sky or by about 80,000 light-years in space. The large cloud is located 160,000 light-years from us and the small cloud is about 180,000 light-years away.

The other important member of the Local Group is M 33, a spiral galaxy seen nearly face on. It is smaller than the Milky Way and M 31, being only 60,000 light-years across, and containing only 4 percent as much matter as the Milky Way galaxy.

Our location inside our own galaxy makes it difficult for us to picture what the Milky Way looks like from the outside, but current opinion holds that someone located in M 31 would probably see a galaxy not very different from what we see of M 31.

The Local Group galaxies appear to be linked in an endless dance through space since they seem to be gravitationally tied together. No member will wander too far off into space without feeling the inexorable pull of the other group members and, therefore, return to the fold.

Other groups of galaxies fill space. Sometimes the galaxies are so closely packed (relatively speaking) that they are called galactic clusters. The largest clusters contain thousands of members. About fifty clusters or groups are located within 45 million light-years of us and they appear to occupy a limited region of space almost as if they are all inside a supergalaxy made up of the visible galaxies. This arrangement is called the supercluster, but the evidence for it is very controversial.

The interaction between galaxies as they pass one another in space, much as a comet might pass the sun, appears to be important in producing some of the spiral patterns seen in galaxies. Such encounters cause distortions in the outer regions of the galaxies and also cause matter to be drawn out in large streamers that thread through the space between the galaxies (see chapter 22).

Searches for material in intergalactic space have been made for many years, without success — until recently. We might believe that such material should exist, much as interstellar matter exists in many galaxies. Stars form out of interstellar matter, and after they have formed the material not used up still fills the voids between the stars. Matter should also be left over after galaxy formation, provided there are similarities between the two processes. But, as was mentioned above, galaxy formation is a little-understood phenomenon and searches for large amounts of intergalactic hydrogen, the basic material expected to exist there (hot or cold), have revealed nothing in the past. Recently, however, clouds of hydrogen have been found between galaxies (see chapters 22 and 24), but they do not uniformly fill space. These clouds are in the form of streamers of matter hundreds of

148 GALAXIES

The beautiful spiral galaxy M 81 showing spiral arms closely wound in its central regions and flaring outward in the outer parts of the galaxy.

thousands of light-years long, and if we can learn why they have this characteristic shape, we might have a clue to galaxy formation.

Ultimately, when all the matter in a galaxy has been used up in star formation, and the stars have run out of hydrogen gas and died, we might expect that the galaxy itself will fade away and remain a hulk containing billions of dead, cold stars. Is it possible that such galaxies already exist? No one knows, since none has ever been found, nor is it likely to be, except by the most indirect means. At present, the thinking about the stage of life or death of a galaxy is very limited. It is interesting to speculate that such galaxies might well be located inside clusters of galaxies. This would be convenient, since many of the larger clusters do not appear to contain enough visible matter to account for their existence as clusters. In other words, because the amount of visible matter in those objects does not produce a gravitational force strong enough to keep the galaxies together in space, why do we see a cluster at all? It should have disintegrated long ago. This missing-mass problem is found in many clusters of galaxies.

CHAPTER 20
THE SHAPE OF THE MILKY WAY

And with an awful dreadful list
Toward other galaxies unknown
Ponderously turns the Milky Way . . .

This insight by Boris Pasternak portrays some of the splendor inherent in the star-disk that is our Milky Way galaxy as it slowly wheels around its hub once every 300 million years. The most traditional view of our Milky Way galaxy, with its 200 billion stars, is that it is much like a flat disk in appearance, about 600 light-years thick and 100,000 light-years across. The sun is about 30,000 light-years from the center which is located in the direction of the constellation Sagittarius (see chapter 21). The size and the shape of the Milky Way are very difficult things to infer since we are unfortunately located inside our galaxy. Were we located just outside it then we would clearly be able to see where the stars, gas, and dust that make up our galaxy are located.

Many distant galaxies have a fairly well-defined spiral pattern, which means that the main concentration of stars and dust follows a spiral outline, starting near the center of the galaxy. Our Milky Way is also thought to be a galaxy that should show such a spiral pattern from afar. The trick is trying to infer from data we collect here on earth, located inside that Milky Way, just what the structure of the Milky Way really is. We can actually see very few of the stars and little of the dust inside our parent galaxy. Dust blinds our sight to any but the nearest stars, nebulae, and dust clouds. Beyond a distance of 5,000 to 10,000 light-years from the sun we see relatively little of the constituents of our galaxy. We encounter a particularly serious problem as we look toward the center of the galaxy since we are also blinded by the enormous numbers of stars located in the star clouds of Sagittarius. Indirect methods have to be used to infer the structure of the Milky Way.

The appearance of the Milky Way's band of stars and dust seen clearly sweeping across the skies at night is a consequence of our galaxy's flattened, disk-like shape. When we look out along the disk we see many stars giving the milky appearance, but when we look at right angles to the disk, we see relatively few stars.

The original estimates of the size of the Milky Way, as well as its shape, were based on the observations of the distances to globular clusters. These old groupings of stars were found to occupy a spherical region of space centered on what we now know is the center of the Milky Way about 30,000 light-years away. Estimates of the thickness of the galactic disk are made by studying how the numbers of stars decrease with distance from the central lane of the Milky Way.

Optical data on the distances to clusters of young stars and to individual stars located within about 10,000 light-years of the sun have allowed an outline to be drawn of the location of most of these young stellar objects. It appears that the sun is perhaps not in a spiral arm, but rather in a spur-like projection (called the Orion spur) from one of the arms that runs near us. In the direction of Perseus we see many young stars and

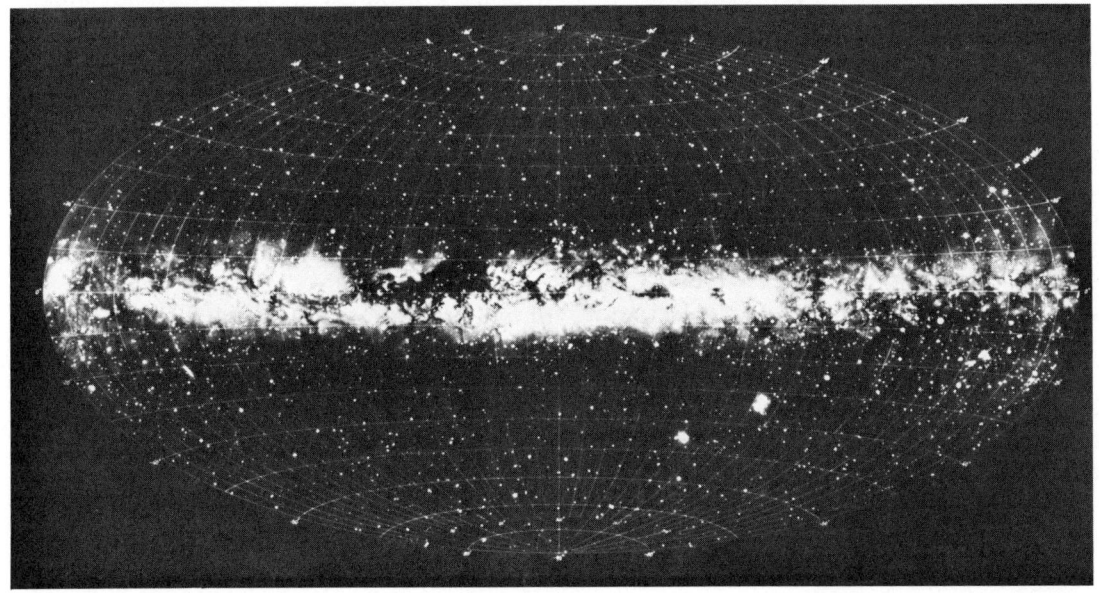

A view of the complete Milky Way produced by two Swedish artists who dotted in 7,000 individual stars and painted the nebulous parts. The sky is shown on a flat projection using a coordinate system based on the Milky Way, and the galactic latitudes and longitudes are drawn in. The top of the picture corresponds to the galactic north pole, the bottom represents the south pole, and a line running through the middle of the Milky Way is the galactic equator.

groups of stars, such as the famous double cluster in Perseus, which all form part of another spiral feature called the Perseus arm. It is located about 10,000 light-years away. Toward the constellation of Cygnus we find a very large number of young stars, gas, and dust clouds, and it is here that we appear to be looking down a section of spiral arm called the Cygnus arm. The Orion spur is probably part of this arm. Toward the galactic center we find another spiral arm sweeping from the region of Sagittarius to Carina. These three major features are the only ones clearly distinguishable optically.

The most recent attempts to formulate a good picture of what the Milky Way galaxy looks like from the outside comes from the work of radio astronomers who study the radio emissions from clouds of hydrogen gas that exist everywhere between the stars. The basic hydrogen atom consists of a proton at its center with an electron orbiting about it. Each of these particles is also thought to spin about an axis of its own, much in the way the earth spins on its axis as it orbits the sun, which in turn spins on its axis. In the cold reaches of space the electron and proton can, however, change their direction of spin at random. When such a spin-flip occurs, the energy that the atom carries changes.

Under the suitable conditions found in the clouds of hydrogen gas between the stars, the hydrogen atoms can radiate a radio signal as a result of this spin-flip. The hydrogen clouds can be observed only in this way since they are otherwise totally invisible. Radio emission from interstellar hydrogen gas was first discovered in 1951; the radio signal has a single wavelength near 21 cm. Since then, many astronomers have been studying this gas, not only to obtain information about the gas clouds themselves, such as temperature, density, and size, but also to learn about the distance to the hydrogen clouds. Based on distance measurements one can locate the hydrogen clouds in three dimensions and draw a picture of what the Milky Way galaxy must look like from outside.

This process is far from perfect, however, since astronomers need to make several important assumptions that allow them to make decisions about where the hydrogen gas is located. Basically, the distance to a cloud observed in some direction can be estimated by measuring how fast it is coming toward us or moving away from us (this is measuring the so-called Doppler shift). If we also know how the Milky Way galaxy is rotating, then we can get a distance to the hydrogen responsible for the signals being observed.

The sun and the region around the sun are moving at about 250 kilometers per second through space about the center of the Milky Way. Since all the gas, stars, and dust near the sun also partake of this motion, we are not easily aware of our motion around the hub of the Milky Way. Someone sitting on the rim of a bicycle wheel would be unaware of motion if he looked at only the rim and spokes of the wheel. If he looked out at the distant pavement, he could discover that he was moving in circles. In the galaxy we need to look at distant galaxies to discover how we are moving, but we can hardly wait for 200 million years (the time it takes to move around one revolution) for us to be sure we are moving in circles. We therefore make the assumption that our motion is such that we do indeed move about the galactic center in a circle and that all other matter in the Milky Way is doing the same; unlike the bicycle wheel analogy, however, it is not rotating as a solid object. Regions closer to the center are moving more slowly than we are.

Assuming circular motion, and using the observed Doppler motion of the hydrogen, we can estimate the distance to the hydrogen gas almost anywhere in the galaxy using simple mathematics. Hydrogen seems to be structured in highly elongated features, like strings or ribbons of matter, spreading out over large distances across the sky. This is what we expect if our galaxy is a spiral galaxy like so many others we see in deep sky photographs.

Recently, it was discovered that in some parts of the sky there appears to be hydrogen gas moving in strange ways that indicate motions other than simple circles around the galactic center are important. There are large numbers of hydrogen clouds apparently coming toward us at up to 200 km/sec in much of the northern sky in areas well away from the Milky Way. These are loosely called "high velocity clouds." In the southern skies there appear to be similar clouds moving away from us. Their presence cannot be explained if we think of the galaxy as a flattened disk rotating in a simple way. They can only be explained if we abandon the idea of circular motion for hydrogen at distances far from the center of our galaxy, even well beyond the point where our sun is located. In addition, we need to question the concept that the galaxy is flat everywhere. If we drop these prejudices, then we can accept the fact that maybe our galaxy is highly distorted in its outer regions and that such distortions explain the existence of the high velocity clouds. This distortion is probably due to an interaction between the Milky Way and the neighbor galaxies, the Magellanic clouds, which appear to have passed by our galaxy hundreds of millions of years ago much as a comet passed by the sun.

We might recall that two thousand years ago we believed the planets also moved in perfect circles, but now we know better than that. It was not until the 17th century, when Kepler suggested that ellipses better described planetary motions, that we finally understood how the solar system was put together. Now it seems that noncircular motions are also playing a role in the motions of high velocity clouds of hydrogen gas in the galaxy.

It should be stressed that the existence of these hydrogen clouds moving in peculiar ways has been a problem for years and the interpretation that they are outer parts of the Milky Way moving in strange ways and distorted by the passage of the Magellanic clouds is only beginning to gain ground as a result of the discovery of similar clouds near other galaxies. New observations by radio astronomers in Australia suggest that galaxies do not wander about as isolated entities as was previously thought. There may be long streamers of hydrogen gas criss-crossing intergalactic space, linking galaxies. It then becomes difficult to decide whether such clouds of matter are really the outer parts of one galaxy (a distant spiral arm section) or whether they should be called intergalactic hydrogen.

Our knowledge of what the Milky Way looks like from the outside is slowly improving and it is clear that, based on the study of the radio emission from hydrogen gas, we are not living in an ordered, flat-disk galaxy. Finally, no one yet knows much about the stars that might exist in the most distant parts of our galaxy. Hydrogen gas in other galaxies is known to extend well beyond the visible galaxy and, in the case of the Milky Way, the extent of the hydrogen may be as much as 150,000 to 200,000 light-years across and the thickness of the Milky Way in the outer parts may be as much as 6,000 light-years.

CHAPTER 21
HIDDEN MYSTERIES OF THE GALACTIC CENTER

Through today's telescopes, astronomers scan the far fringes of the visible universe, watching events occurring billions of light-years away. Meanwhile, the center of our own pinwheel-shaped galaxy, the Milky Way, remains one of the most mysterious and uncharted regions of space. One major problem is that the sun, along with 250 billion other stars, dwells within the Milky Way's broad disk. Vast clouds of obscuring interstellar dust drastically reduce optical telescopes' observational range (within the galactic plane) to a fraction of the Milky Way's 100,000 light-year diameter. This situation can be compared to standing in a suburb on a foggy night and trying to locate and sketch the center of the adjacent large city.

Clues about the nature of the galactic nucleus must be gathered in parts of the electromagnetic spectrum unaffected by dust — infrared, radio waves, and X rays. Two of these regions in particular — infrared and radio waves shorter than one meter — are allowing astronomers to peek into dynamic processes and unusual structures at the Milky Way's heart. Unraveling the shape and behavior of the nucleus is vital to understanding our galaxy's evolution. New insights into violent events detected in other galaxies could be an added bonus. Observations of radio galaxies, Seyfert galaxies, and possibly quasars reveal that galactic nuclei are sometimes the focus of titanic explosions, overwhelming in their energy output and scale of disruption.

In exploring the Milky Way's heart, the first problem was encountered in pinning down its exact location. Two simple observations pointed the way decades ago. Globular clusters, the oldest known collections of stars accompanying our galaxy, are distrib-

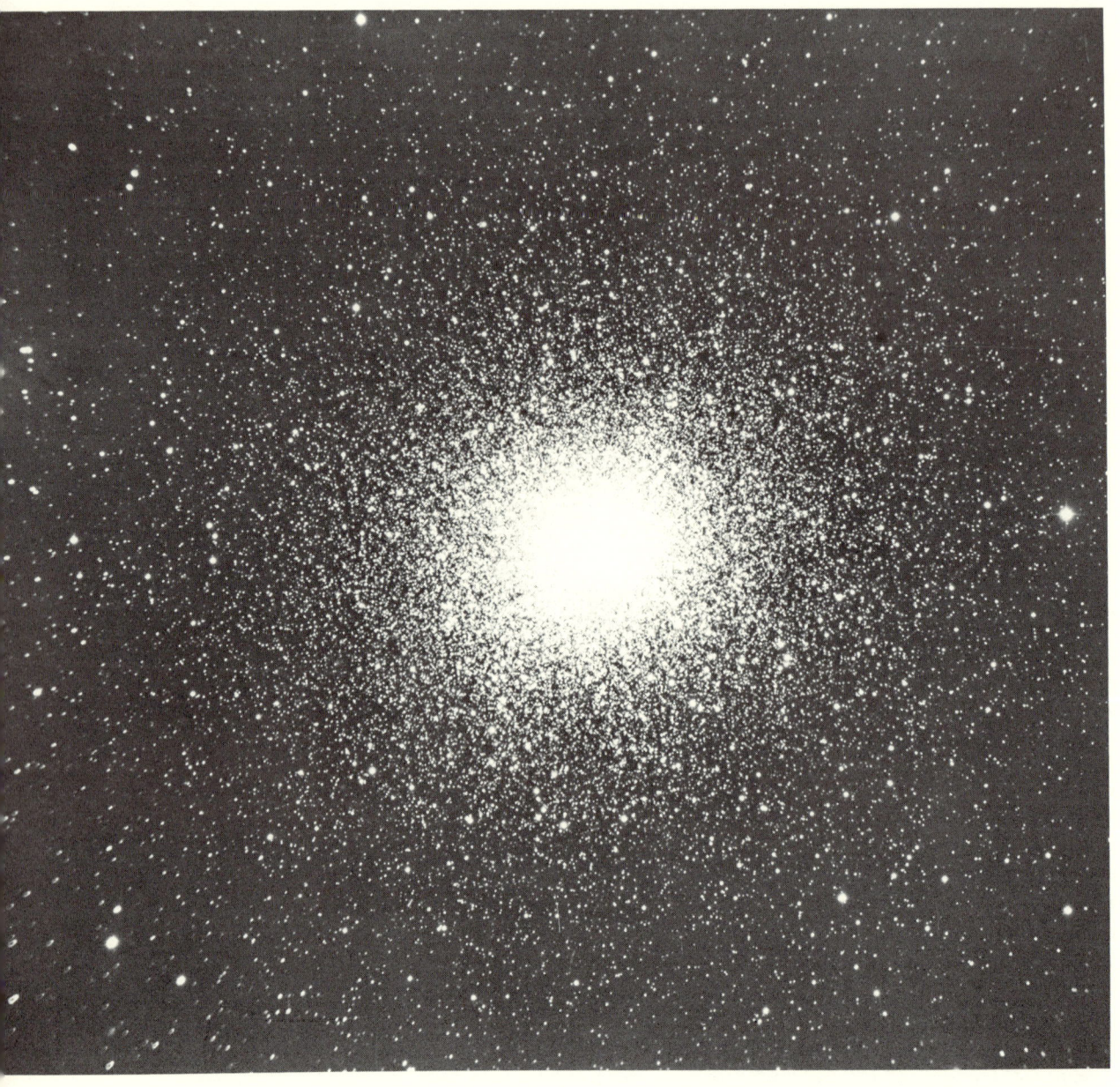

The globular star cluster M 13, containing tens of thousands of stars in a region of space nearly 100 light-years across. All these stars were born out of the same cloud of matter billions of years ago.

uted in a spherical region of space bisected by the Milky Way's disk. The center of this sphere of globular clusters, and hence the center of the galaxy, is located in the direction of the constellation Sagittarius. Early estimates placed the galactic center at a distance of 30,000 light-years from the sun. Currently there is talk of reducing that value to 27,000 light-years.

During the last few years, observations of radio signals from the densest parts of the Milky Way have allowed astronomers to zero in on the galaxy's center. Radio maps of the sky show a very strong radio source surrounded by several other weaker radio sources in the direction of Sagittarius, approximately where the center should be located according to the globular cluster data. Radio data now pinpoints this central source to the incredible accuracy of a few seconds of arc and measures its diameter as being only about 2 light-years.

Confirming evidence that this is the central source comes from a considerable amount of infrared radiation emitted from the same location — infrared radiation so intense that it cannot be easily interpreted unless we believe that something very special is occurring there. Present explanations for the infrared hot spot propose that a very dense cluster of stars at the center is surrounded by a thick cloud of dust absorbing the light from the stars and reradiating it as heat. Having a diameter of a few light-years, this cluster contains the equivalent of 100 million sunlike stars. Nevertheless, absolutely no light escapes through the enshrouding thick cloud.

More infrared radiation surrounding this very dense dust core indicates the presence of additional dust and stars. Apparently a large number of stars surrounding the

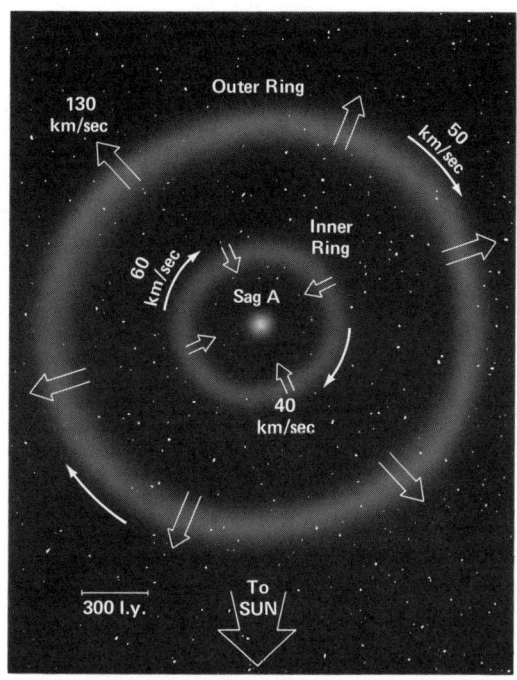

A schematic representation of the motions of gas and stars near the galactic center.

nucleus cluster — occupying a region about 100 light-years across — are also cloaked by a dense dust shell. Only one trillionth of the light emitted by those stars escapes outward through the dust.

The strongest and smallest diameter radio source found in the galactic center seems to coincide exactly with the bright infrared core. The type of radio emission associated with that source (known as Sagittarius A) is produced by high energy electrons spiraling about magnetic field lines, and is usually associated with exploding stars or galaxies. This is called a nonthermal source, as opposed to a thermal source, like the hot gases found in the Orion nebula (one of many emission nebulae).

Although unusual, a nonthermal radio source in a region where there must be an enormous number of stars present is not entirely unexpected since such sources are found at the center of other galaxies. Having an incredibly large number of stars packed in a small volume of space apparently makes conditions favorable for matter to emit such radio signals.

In addition to Sagittarius A, there are several thermal emitters — large clouds of hot gas, lying near the galactic center. Many young, hot stars heat the gas clouds, which in turn reradiate their energy as radio waves. The center of our galaxy is dotted with about a half dozen of these very large emission nebulae — totally invisible to us, but detectable because of their strong radio signals. The most remarkable, Sagittarius B2, has a companion cloud that contains many of the molecules recently discovered in space.

Radio observations of interstellar molecules show that some molecules such as formaldehyde, carbon monoxide, and hydroxyl are spread throughout the galactic center region. The direction and distance of the emitting regions provides a clearer picture of what is happening in the nuclear regions of the galaxy just outside the inner concentration of stars and dust.

Apparently there are two doughnut-shaped dust clouds ringing the galactic center. An outer ring of molecular material, probably consisting of many individual clouds, has a 1,600 light-year diameter. Containing the equivalent of 100 million suns, this ring is expanding at 130 kilometers per second (km/sec) and rotating at 50 km/sec. Inside that main molecular ring is an 800 light-year diameter ring, containing hydroxyl and formaldehyde. Also rotating at 50 km/sec, this inner ring is falling in toward the galactic center at 40 km/sec.

The expanding ring could have been produced by an explosion and the inner ring is material affected by the outward moving shock wave. Such shock waves could possibly trigger the formation of stars. Vast quantities of molecules may be created in the clouds of gas and dust trapped within the shock wave.

Observations of the radio signals at 21 cm wavelength (the characteristic signal of interstellar hydrogen) show that these molecular doughnuts lie inside a rapidly rotating disk of hydrogen gas. Some regions of this hydrogen are also expanding away from the galaxy's center, implying that an explosive event — occurring about 10 million years ago — initially ejected matter outward at 2 million km/hr. Today astronomers are puzzled as to why this disk of matter still exists following such an incredible explosion, and they are even more mystified about what initially triggered the blast.

Moving farther outward from the center, additional evidence for the explosion comes from observing the spiral arm structure of our galaxy in the region just outside the galactic nucleus. Radio data reveal a distinct spiral arm on our side of the galactic center moving away from the center at 52 km/sec, and another arm on the far side of the center moving outward at 70 km/sec. An explosion initially ejecting two large blobs of matter in nearly opposite directions (at a small angle to the Milky Way so as not to destroy the central disk) can explain the motions. These blobs were subsequently drawn out to form two spiral arm-like structures. This structuring is similar to the appearance of many radio galaxies that are double sources.

There are also a number of X-ray sources in the direction of the galactic center, but none have been located accurately enough to suggest that they are precisely at the center of the Milky Way.

There has been speculation that a giant black hole — fueled by the nucleus' high density of stars — might be the power plant behind such a colossal explosion. Although observations do not rule out such a phenomenon, a black hole of any size is not required to explain the present data.

Uncovering what is happening at the Milky Way's center is one of modern astronomy's most exciting detective adventures. At last we are beginning to fit together many tenuous pieces of evidence into a picture of the hub of our island universe. No doubt this present portrait will be changing in the coming years. But it is unlikely that any new surprises lying just around the corner will force us into rejecting the basic concepts of today's models of the galactic center.

CHAPTER 22
INTERACTING GALAXIES

Our nearest extragalactic neighbors are the clouds of Magellan, two irregular galaxies about 175,000 light-years from us. On a clear night in the southern hemisphere, a night so clear that the Southern Cross is lost in the vivid band of the Milky Way itself, you can see both clouds with the naked eye in the constellation Dorado.

These two relatively small galaxies (30,000 light-years across) orbit about our own Milky Way galaxy in much the same way that a comet orbits the sun. At the same time they appear to have a devastating effect on our galaxy when they pass close to us.

There are many known cases of interactions between galaxies. Many interacting galaxies are seen in the Arp Atlas of Peculiar Galaxies. They often appear as two or three separate galaxies in a photograph, but with obvious bridges of material between them. At left, we see a black and white photograph, actually a negative print, of Arp 295, an interacting galaxy pair.

The long sweep of emission between the two galaxies is an intergalactic bridge that streams through 400,000 light-years of space. Astronomers believe this bridge was produced when the two galaxies swept near to one another many hundreds of millions of years ago when the mutual gravity of the galaxies dragged out matter from their disk-like structures. The photograph on page 162 shows NGC 4675 (known as "The Mice") and the intergalactic arches that join them and radiate from them.

We know that such intergalactic bridges are a norm of the universe, according to some recent theoretical work done by Alar and Juri Toomre. These scientists used a computer to simulate the effect of one galaxy on another when the two pass close to each other in space. Their original intention was

The interacting galaxies known as Arp 295, showing long streamers of matter dragged out of one by the other as they passed close to each other at some time long ago.

162   GALAXIES

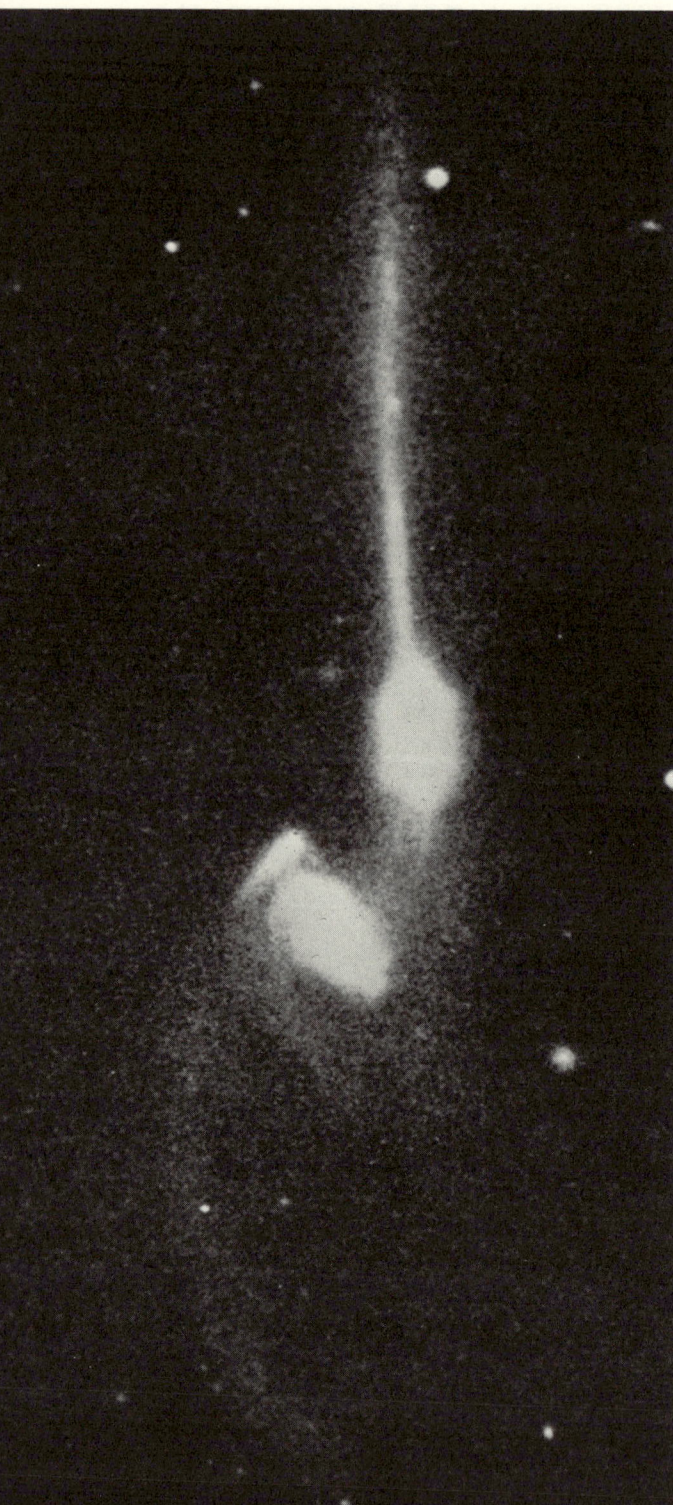

The peculiar galaxies known as NGC 4675 or "The Mice." From the distortion of their outer regions, we can clearly see that they are interacting with one another.

to examine what effect the Magellanic clouds would have had on our galaxy, if the clouds had passed close to us. In the computer they used two disk-shaped objects to represent our galaxy and the large Magellanic cloud. Both disks consisted of many separate points that simulated the existence of matter (stars) in the galaxies. Then using complex mathematics, they sent one disk past the other in three-dimensional space to see what would happen. The results were surprising.

The original motivation for doing the computer experiment was to explain the fact that our galaxy appears to be shaped somewhat like a hatbrim. One edge of the brim bends up, the other down. This is not quite the traditional view presented by standard astronomy texts, where our galaxy is always described as being a flat disk with a bulge at the hub. The edge of our galaxy, in the direction of the constellations Cygnus and Cassiopeia, bends to the north of the Milky Way. Matter in these distant regions is very patchy and seems to move with strange velocities. The Magellanic clouds were passed close to our galaxy in various orbits, like a comet passing close to the sun, and they searched for a suitable orbit that might explain the bending of our galactic disk.

They found that if the Magellanic clouds passed by our galaxy in a direction opposite to the way our galaxy rotates, then the clouds would indeed interact with matter in our galaxy and cause its outermost parts to distort and take on the distorted hatbrim

INTERACTING GALAXIES   163

Two interacting galaxies. The large one is a clearly defined spiral known as M 51 and its smaller companion, located a short distance behind it, is known as NGC 5195. The interaction between these two galaxies as they passed one another in space may have generated the spiral pattern we now see in M 51.

shape. They were further able to show that the Magellanic clouds must have come toward our galaxy from the northern part of the sky and crossed the Milky Way plane near the region of the constellation Perseus. At their nearest approach the clouds must have passed the plane of the Milky Way at a distance of about 40,000 light-years from the sun, or about 60,000 light-years from the center of our galaxy. We now see the clouds in the southern sky traveling away from us, perhaps to return to their original positions in about 3 billion years.

But the real surprise was something else. Although the Toomre brothers started with two uniform disks that bent up at the edges during the time of closest approach, the points representing stars in the galaxies started to form a spiral pattern. They went on to model the situation of M 51 (see p. 163) and its companion and found that the outer spiral structure seen in M 51 was probably produced by the interaction with NGC 5195, its companion galaxy.

We know that our sun is located in a galaxy that also has a spiral pattern, although it is difficult to form an accurate picture of what that pattern is (chapter 20), since we are unfortunately situated inside it. Theoretical astrophysicists recently proved that spiral structure could be maintained for many galactic rotations. However, they could not explain the origin of spiral patterns. It appears possible that at least outer spiral structure might be the direct result of close interactions between galaxies in groups or clusters.

We might speculate that after the effect of the interaction with the Magellanic clouds has worn off, it is quite possible that our galaxy will lose its spiral pattern. Stars and gas will slowly collapse inward and form an amorphous disk-shaped object until the next time the clouds come by to generate spiral structure all over again. In addition, if the clouds had been in a direct orbit, they would have caused so much disruption of the Milky Way system that our galaxy would appear totally different from the way it now appears. There are other galaxies that probably interacted destructively in this way, explaining some of the quite remarkable structures we see in photographs of extragalactic nebulae.

If the Magellanic clouds have passed by our galaxy in the way we think, then it is probable that they also dragged out matter from the outskirts of our galaxy and vice versa. But where is the expected subsequent intergalactic bridge? Any good theory ought to predict where the bridge should be — although it would be invisible since it probably contains mainly neutral hydrogen gas. However, radio astronomers studying radiation from this gas should be able to find it. Searches were made in areas where the Toomres predicted the intergalactic bridge might be, but nothing was found.

The story does not end there however. A possible bridge of matter joining our galaxy to the Magellanic clouds may have been found — a very definite "tail" or stream of matter radiating from the clouds has been discovered by Don Mathewson.

Let's look at this newly found stream in the context of its discovery.

One of the most enigmatic problems that confronted radio astronomers for a decade was the existence of hydrogen gas moving at large velocities well above and below the Milky Way (see chapter 20). We expected only to find slow-moving, nearby hydrogen in these directions since the galaxy is supposedly a flat disk. However, if our galaxy is distorted, then the hydrogen in distant spiral arms would appear well away from the Milky Way band and this matter would appear to have high velocities toward or away from us due to the rotation of the galaxy.

One high velocity cloud was found near the south galactic pole — a direction in which distant spiral arms were hardly likely to exist. The south pole cloud, as far as northern hemisphere observers could ascertain, was about 70 degrees long and the velocity of the hydrogen in the cloud ranged from near zero with respect to us to a velocity along its length of about 400 kilometers per second toward us.

Recently, the cloud was found to reach all the way to the Magellanic clouds themselves, where radio emission blended with emission from the Magellanic clouds. The two Magellanic clouds were already known to be linked by their own bridge of hydrogen gas. Mathewson refers to the south pole cloud as a stream of matter. It appears to originate at the clouds and track off into space, if the motions in the stream are interpreted correctly. The patches of clouds seen near and above the Milky Way at longitude 300 degrees may be part of a bridge that is the outer spiral arm of our galaxy reaching to the Magellanic clouds. The stream, however, may be very similar to the structure seen in the illustrations on pages 161 and 162.

CHAPTER 23
EXPLODING GALAXIES

Since we are located in a relatively quiet galaxy known locally as the Milky Way, we are able to live on our planet and study the universe around us at our leisure. (At least, so far we have been able to do so!) But this is not so for any unfortunate beings living on a planet in a galaxy that undergoes an enormous explosion.

In their examination of distant galaxies, astronomers have found many that have actually undergone explosions on a galactic scale. Some of these explosions have torn entire galaxies apart. In other instances observations show two enormous blobs of matter streaming away from a central explosion. In most cases observed so far, the explosion occurred in the center of the galaxies, and often two clouds of energetic particles are thrown out in opposite directions giving rise to the now well-known phenomenon of double radio sources. These exploding galaxies are usually observed by the radio emissions they generate in the two ejected clouds of matter, rather than by a search of sky photographs.

Some exploding galaxies can easily be seen optically. The most well-known one is the Galaxy M 82 in the constellation Ursa Major. This one has not evolved to the characteristic double configuration, perhaps because not enough time has passed. The photograph on page 167 shows M 82 and the filamentary matter radiating away from the main disk of stars and dust. This material is moving at high velocities, many thousands of kilometers per second.

The elliptical galaxy M 87, in Virgo (see p. 144), is an example of a galaxy that has exploded but is apparently only ejecting material in one direction. Violent, chaotic events have generated strong magnetic fields and highly energetic particles in

The Galaxy M 82, showing filamentary matter extending outward for up to 25,000 light-years on either side of the galaxy. This galaxy is also a strong emitter of radio signals; for this reason, it is believed to have undergone a violent explosion of some sort, although it is not proven that the filaments are a manifestation of such an explosion.

and around M 87, making it a "bright" object in radio telescopes. Traveling at almost the speed of light, these energetic particles encounter the magnetic fields and spiral about the magnetic field lines. In doing so, they radiate away some of their energy in the form of radio and light signals. This makes them detectable on earth as a distant source of radio emission.

M 87 is one of a class of objects known as radio galaxies (as is M 82). The central part of this galaxy (see p. 168) is seen to have a luminous jet radiating from it. On page 174, the photo of quasar 3C273 also shows a jet of matter projecting from it. While the quasar is at a distance of about 1½ billion light-years, the photographs show that these two objects are superficially similar. However, the quasar is 100,000 times more luminous than M 87, so it is very clearly a different type of object. Yet the two, when viewed in comparison, look similar. Possibly, similar processes occurred in these two objects to produce the enormous explosions required to send the jets of matter out into space. The jet in quasar 3C273 is about 150,000 light-years long, and the one in M 87 is only 3,000 light-years long. The similarities and differences between this radio galaxy and quasar might give clues about the true nature of quasars.

Explosions in distant galaxies and quasars are quite harmless to us since the cosmic rays produced there will never reach us. We only pick up their highly diluted radio and light signals with our largest telescopes. However, if such an explosion were to occur in our galaxy it would probably wipe out life on our planet — the enormous excess dosage of cosmic-ray particles (high energy protons and electrons) would be fatal. It is unknown whether such a major explosion will ever occur in the Milky Way. We hope our galaxy is not susceptible to an explosion on such a large scale.

Many radio galaxies are elliptical galaxies — usually the largest elliptical in a cluster of galaxies. Our galaxy is a spiral galaxy unlikely to become a radio galaxy. Nevertheless, the nucleus of the Milky Way system does have many of the characteristics of a very small elliptical galaxy in terms of its stellar content. A study of gas motions toward Sagittarius — the direction of the center of the Milky Way — has provided evidence that a small explosion may have occurred at some time in the past in the center of our galaxy (chapter 21). This explosion was not strong enough to influence us

A short time-exposure photo of the galaxy M 87 clearly reveals the jet of matter emerging from its nucleus. The jet has structure within it and both the nucleus and the jet emit radio signals as well as shining by their own light, as seen in this negative print.

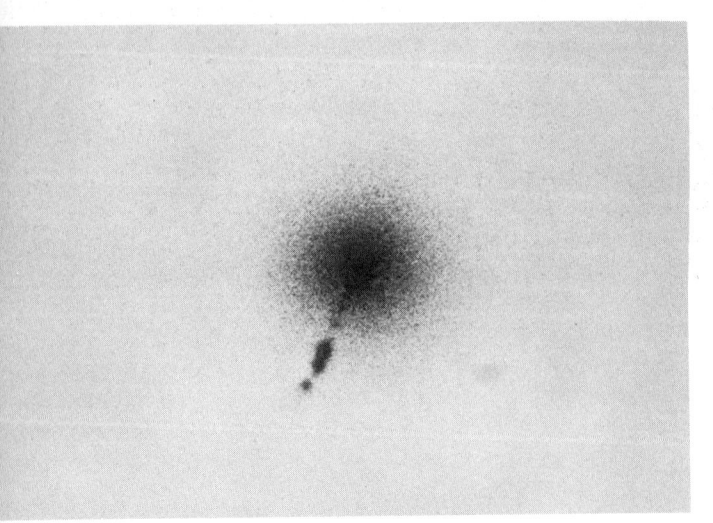

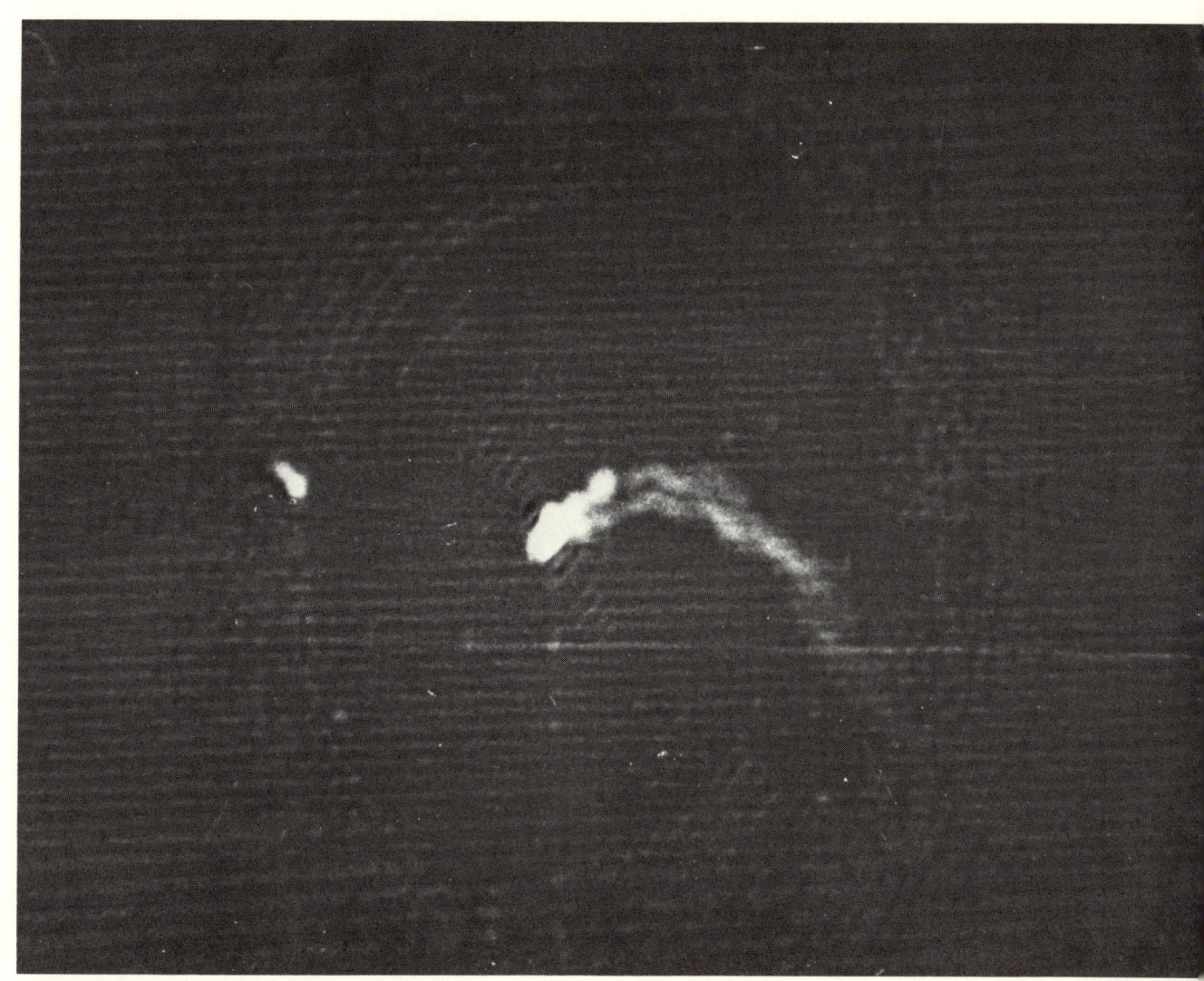

A radio "photograph" of the radio galaxy 3C129, showing a tail of radio emission stretching behind the galaxy as it moves through space. At the head of this radio source a visible galaxy can be seen in photographs. A smaller companion radio source is also seen in this picture. The fact that this tail does stream behind this galaxy suggests that the space between galaxies here contain matter and that the tail is much like the wake left behind by a ship.

at our distance from the center (30,000 light-years).

Recently, another fascinating form of exploding galaxies has been studied by astronomers. This is known as a head-tail galaxy. The first was discovered by radio astronomers in Cambridge, England, and later studied in detail in Holland.

A radio photograph of a head-tail radio galaxy with a smaller companion radio galaxy is shown above. This is actually a photograph of a map produced by displaying the intensity of the received radio signals on a cathode ray tube where the brightness is a measure of the intensity of the received radio signals. This "radio graph," as it is called, was made at the new radio observatory at Westerbork, Netherlands and shows the pair of radio sources known as 3C129 and 3C129.1. The main component, 3C129, shows a long radio tail or stream stretching out from the main nucleus. An optical photograph of this part of the sky shows a galaxy located at the tip of the head of the radio source.

It is thought that the galaxy underwent an explosion that ejected two large clouds of matter into space, again in two opposite directions, similar to the more well-known double radio sources. However, in this case the galaxy was moving very rapidly through space, at several thousand kilometers per second. In addition, there must be material in intergalactic space around 3C129 that acts like a ghostly cosmic wind blowing past the rapidly moving galaxy. As the two blobs of material move out into the surrounding space, the wind sweeps them back in what could be called the wake of the galaxy. In addition, it is thought that the parent galaxy is also rotating so that the two ejected clouds are twisted into a helical shape almost like two corkscrews stretching out behind the galaxy. This explains the double and twisted structure in the tail. The other smaller object is probably a similar, but weaker, radio galaxy in the same cluster of galaxies of which 3C129 is a member.

The curved tail of 3C129 appears to be over one million light-years along. An estimate of the age of this object suggests that the explosion started 200 million years ago.

The enormity of these galactic explosions is quite beyond comprehension — their violence defies comparison. But we can feel secure in the knowledge that we exist on a planet that is a citizen of a normal galaxy.

SECTION VII
QUASARS AND THE UNIVERSE

Beyond the galaxies lie the quasars. What are they? Do they give us clues about the origin of the universe? Did the universe even have an origin? How do astronomers test the basic physics at the root of understanding the universe? As we move outward in space we find many unanswered questions that make the study of astronomy much more fascinating. The existence of quasars strongly suggests that we are dealing with phenomena that present-day physics is at a loss to explain. Perhaps we are making fundamentally wrong interpretations of some data or it might indicate that there are laws of physics about which we know nothing yet. Let's travel out beyond the galaxies and examine the problems.

CHAPTER 24
QUASARS

What is a quasar? No one knows! Of all the discoveries of the last fifteen years, quasars are still one of the most enigmatic because their nature is not yet understood. Are these objects at the very edges of the visible universe or are they relatively close? Are they parts of the normal evolutionary sequence that galaxies undergo or are they unrelated to galaxies? Where does their energy come from that makes them such strong light and radio beacons? Is it possible that the solution to the quasar mystery will involve a fundamental rethinking of the basic physics to which we have been growing accustomed since Albert Einstein's time?

The story of quasars started in the late 1950's when radio telescopes picked up radio signals from many small point-like objects spread over the sky. Advances in radio technology meant that fainter and fainter radio sources of emission were being detected. Some of these small radio sources appeared to be located at the same position on the sky occupied by star-like objects. One of these star-like objects was accurately located when the moon crossed in front of it. When the moon covers a radio source or a star in this way, the phenomenon is called an occultation. By monitoring the radio signals from the source and watching the moon occult the source, the accurate location of the object can be found by timing its disappearance behind the moon and its later reappearance.

In actual practice, two passages of the moon are needed to unambiguously pinpoint the radio source position. In 1963, two such occultations occurred for the radio source known as 3C273. The observations were made in Australia and the analysis of the position of this source showed first that it was a double radio source, with the two components separated by about 20 seconds of arc. Subsequent photographs of that spot revealed the presence of a star-like object with a jet of luminous matter protruding from it. The method of lunar occultations has since been applied to finding the accurate positions of hundreds of other radio sources. Since the position of the moon at any time is accurately known, the time at which the source disappears behind it (or re-emerges) can easily be converted into an accurate position for the limb (edge) of the moon and hence for the radio source. The inaccuracies in the position determined for a strong radio source depend on how well the shape of the moon is known.

Thus, it appeared that 3C273 was a star, except for the unusual jet issuing from it. At about the same time, the accurate positions for several other star-like radio sources had been found using different techniques and the rumor spread that some stars in our galaxy were also strong radio emitters, but for reasons that were not immediately obvious. (The sun emits radio signals that would not be detectable on earth if it were located many light-years away from us.)

The puzzle of the radio stars was studied by many astronomers and, in 1963, Marten Schmidt at the Hale Observatories obtained a sensitive optical spectrum of the radio star 3C48. The spectral lines he found did not match those of any known star, and it was not until he tried to see whether the lines were shifted in wavelength that he made the important breakthrough. The spectrum of 3C48 could best be explained if he assumed that it showed a considerable redshift (that is, the spectral lines expected for a particular atom were shifted to longer, or redder wavelengths). Changes in redshift are well known for galaxies (chapter 13), but have not been observed for stars. More distant galaxies are known to show a larger redshift. If we interpreted the apparent redshift of 3C48 in the same way as we did for galaxies, then 3C48 was certainly not a star; furthermore it was located well outside our Milky Way galaxy. Redshifts of galaxies are generally used as distance indicators, and in the case of 3C48 it appeared to be located 3 billion light-years away, well beyond the most distant galaxy previously observed.

Redshift measurements of several other radio "stars" quickly followed and showed that they all appeared to be located at very great distances from us. These objects were clearly not stars, and they were named "quasi-stellar radio sources" for a while. That name was quickly abbreviated to quasar. (Many astronomers predicted that such a name would never stick.) The largest quasar redshift implies the quasar is traveling away from us at 80 percent of the speed of light, placing it 15 billion light-years away.

Astrophysicists now confronted a fascinating problem. Here was a class of radio emitters visible with optical instruments, but that were located well beyond any other objects previously studied. The distances to some quasars were thought to be more than 10 billion light-years. Nevertheless, we on earth were picking up a lot of light and radio energy from these objects, which meant the sources had to be under-

Photographs of several quasars. Their apparently star-like images reveal little about their nature when studied more carefully. The quasar 3C273 has a jet protruding about 300,000 light-years from it, apparently streaming away from the quasar's visible center.

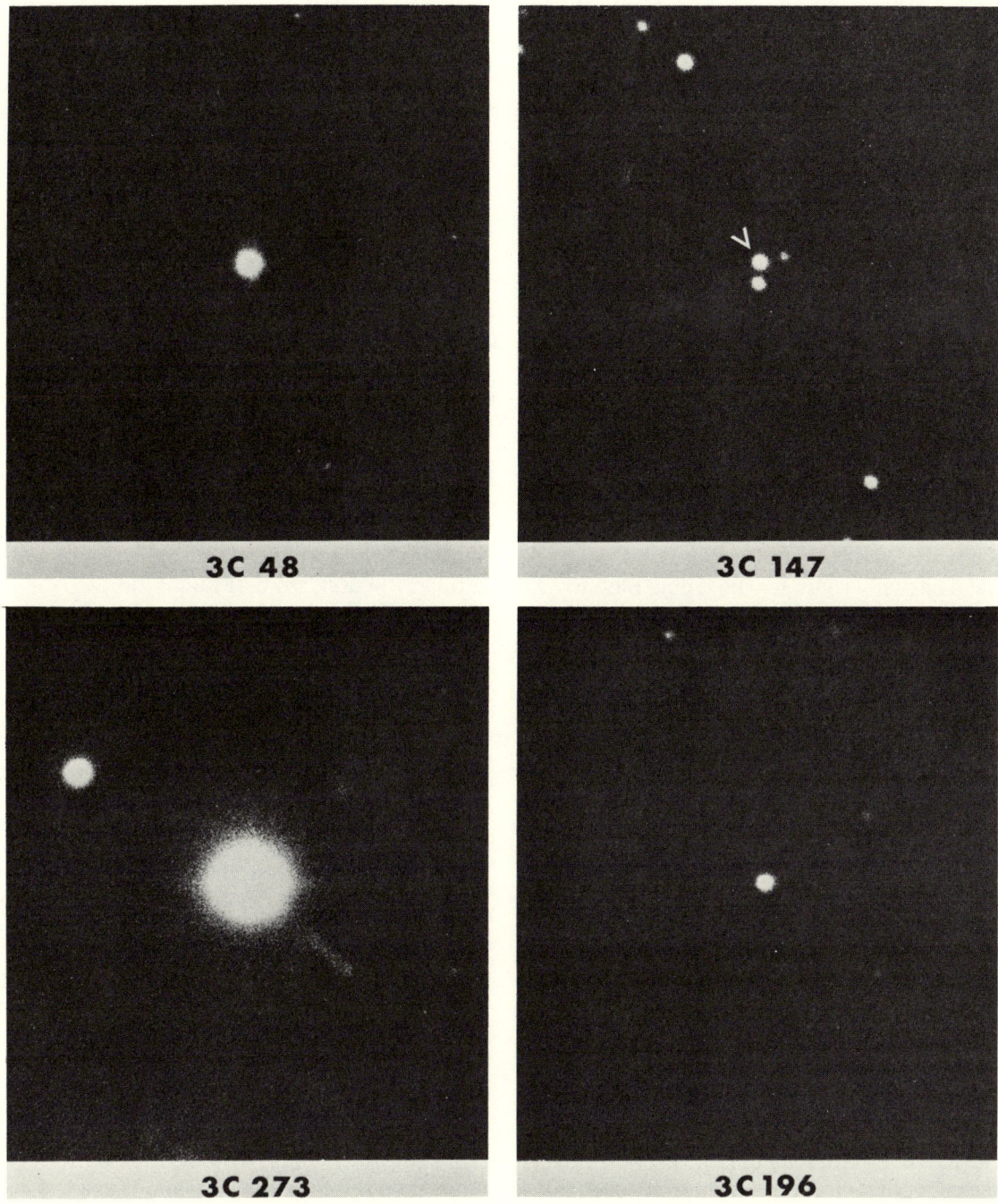

going some very energetic internal events in order to be detectable over the enormous distances. Ordinary galaxies are simply not seen that far from us. No process was known that could account for such vast quantities of energy being produced on a galactic scale.

The problem was further compounded when quasar 3C273 showed variations in light emission from year to year. This was discovered by examining its image on photographic plates extending back over eighty years. If a source of light is changing on a time scale of one year it has to be less than one light-year in size. Consider that some large cloud of matter is capable of emitting radiation as the result of some triggered process within it. The impulse from the trigger will, at most, travel at the speed of light across the cloud. It would therefore take as many years for the whole blob to start shining as it would take light to travel from one side to the other. In reality, such a triggering impulse might well travel at much less than the speed of light, so that if we actually observe a change of light output in one year, the size of the object that we see is probably much less than one light-year across.

This phenomenon compounded the quasar problem because the energetic source of emission inside the quasar varied from year to year and, therefore, had to be confined to a very small volume of space less than a light-year across. There appeared to be no way, using the known laws of physics, to explain the emission of this amount of radiation from so small a volume. The problem is still with us. Since those days in the mid-1960's, quasars have even varied on time scales of months, and very sensitive angular measurements have shown that there are tiny spots less than a light-year across within the quasars that appear to be the sources of all radiation emission.

One way out of the energy problem is to consider that the quasars are much closer than we expect from their redshifts. If this is so, the energy problem is lessened, but then we have to explain the redshifts in some other way. In any event, we still do not know what a quasar is, nor do we know what causes the redshift. A redshift can also be obtained when light escapes from a very massive object. This is called the gravitational redshift. But it does not seem to account for quasar redshifts, even in part, since some of the recently measured redshifts are simply too great.

A considerable amount of effort is being put into the search for faint optical emission around the quasars that might indicate they are parts of galaxies — perhaps a bright central core surrounded by a galaxy. So far, little, if any, success has been obtained in these searches. Also, if quasars are a strange form of distant galaxies, we might expect to find other galaxies around them, since most galaxies occur in groups or clusters. So far, no conclusive proof for the existence of quasars in clusters of galaxies has been found.

There is another class of radio emitters, the so-called radio galaxies, which are nearby visible galaxies that are strong radio sources. Some of them are as energetic as the least luminous quasars; if it were not for the fact that we can readily photograph the radio galaxies and recognize them as galaxies, there are no other obvious properties that distinguish them from quasars. This naturally leads to much speculation about the relationship between these two classes of objects; perhaps quasars are simply radio galaxies with very bright cores located much further away. At the time this book was being written, a radio source known as 3C123 had been photographed with very specialized electronic techniques; it appeared to be a radio galaxy located 8 billion light-years away. This is much further than any galaxy has ever been seen and is a distance at which only quasars have been discovered until now. Clearly, observations of such distant galaxies will have an important bearing on our future understanding of both radio galaxies and quasars.

The quasar enigma has recently been worsened by further discoveries of strange happenings within them. A few years ago there were individual parts of a quasar that appeared to be traveling away from each other at speeds greater than that of light. Since Einstein's laws of relativity state that matter belonging to our physical universe cannot travel faster than light, this discovery caused quite a furor.

In order to better understand the faster-than-light story we will first concentrate on the quasar 3C273. As mentioned above, the radio signals from 3C273 are coming from the core of the star-like object as well as from the jet itself, the two being separated by about 20 seconds of arc. Since the distance to 3C273 is thought to be about 3 billion light-years (based on its redshift), the calculated length of the jet is about 300,000 light-years. In recent years, closer looks at 3C273 have shown that the source of radio signals inside the quasar itself was in fact more complex than previously thought. The core of 3C273 was found to contain at least two small sources, each a light-year or so across and separated by about the same amount of space. Since the original radio observations were not sufficiently accurate, radio astronomers could not have been certain that the core source was not more complicated, consisting of, perhaps, three or even four parts. However, if a simple double source was assumed, the radio measurements made over the space of about six months seemed to show that the two members had apparently moved apart. Since the distance to the quasar was known, it was possible to calculate how far the two central "blobs" had traveled apart in six months. This meant that the actual speed with which the two components were flying apart could be calculated. It turned out that this speed was greater than the speed of light! How was this possible? According to our known laws of physics, it was impossible.

Breaking well-known laws of physics could be avoided since there are other ways we could get the illusion that parts of the quasar were traveling faster than light. For example, we could imagine the presence of several parts of the radio source, each flashing on and off independently. They could be lined up in such a way that the flashing of one part after another would lead us to think that we were seeing a single one traveling very fast.

This is similar to the problem of the radio data since the observations were spread six months apart and no one knows what happened in between. Did the sources actually move apart or did one fade and another start to shine at another point in the quasar? The only way to answer that question is to make more careful observations of the sources.

178   QUASARS AND THE UNIVERSE

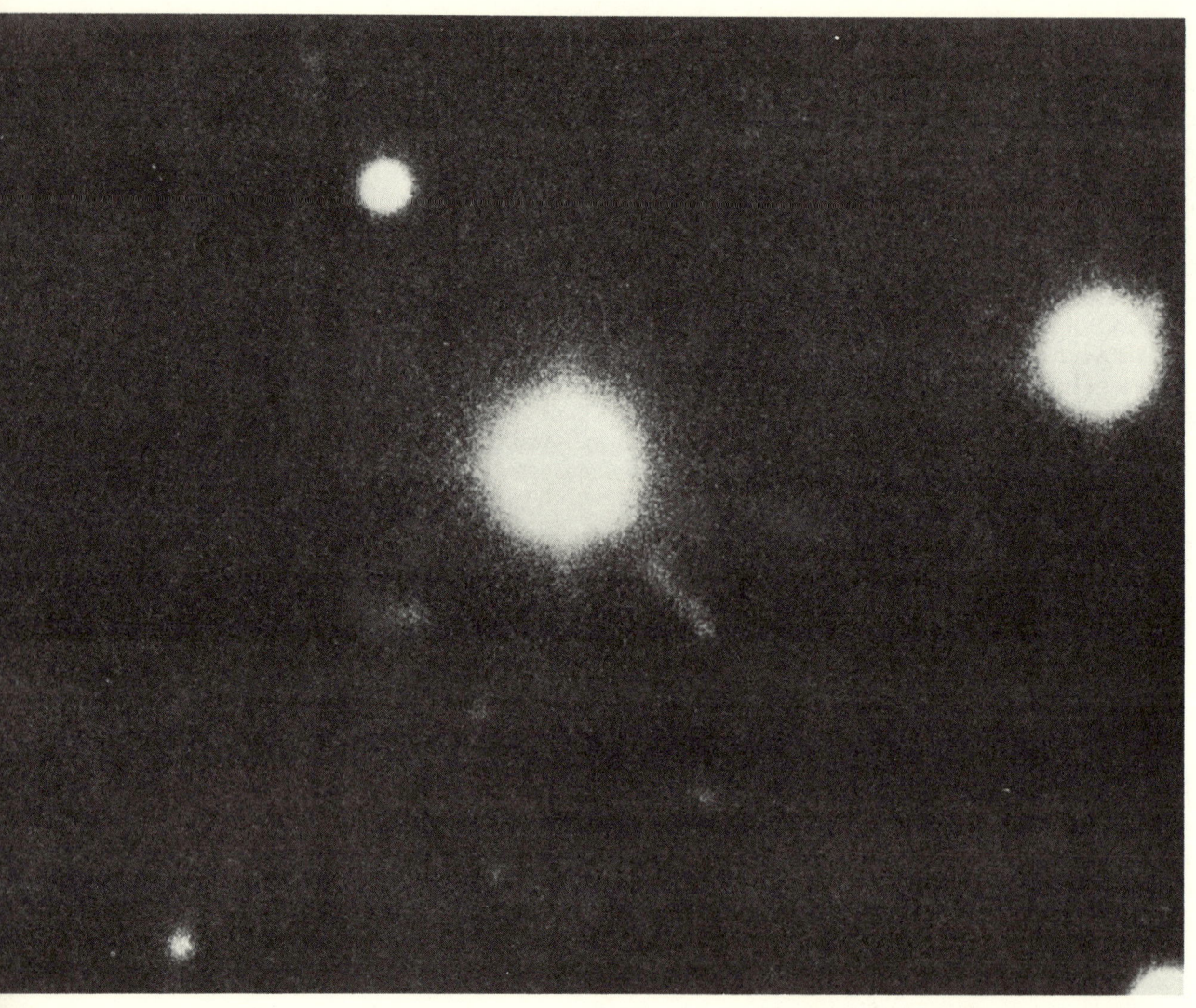

A greatly enlarged view of the quasar 3C273 showing the jet of matter protruding from it. The other objects in the photo are stars in our galaxy and show that, apart from the jet, there would be no way to tell the quasar from the stars. Such similarity led to the original idea that these radio sources were in fact stars, until it was discovered that they are objects way outside our galaxy, perhaps beyond the most distant galaxies we know about.

This has been done and the data compiled are interesting.

Using three radio telescopes working in combination, one in California, one in West Virginia, and one in Texas, the quasar 3C273 was much more carefully watched over a period of another year. The new observations cleared up some minor problems and raised other major ones. First of all, the core source in 3C273 was found to contain not two, but three small sources in a line. The previously suspected double was an oversimplified interpretation of the older data. However, this did not explain the faster-than-light problem, since now the triple source appears to be moving apart! The simplest interpretation of the motion indicates that the outer two members of the triple source are each traveling at five times the speed of light away from the center of the quasar!

The separation between the parts of the triple source is something like a few thousandths of a second of arc and the rate at which they move apart as seen against the background of the sky is only tenths of a thousandth of a second of arc per year! At the distance of 3C273 these individual components are only a few light-years apart, but in another year they should be about 10 light-years apart if the observed effect does indicate faster-than-light travel. No one yet accepts this explanation, however, since there are now several other possible explanations for the phenomenon that do not require our acceptance of faster-than-light travel. The first alternative explanation forces us to question the distance to the quasar. If the quasar is, say, ten times closer than we thought it was, then the speeds involved are all ten times less since the sizes that we attribute to the quasar are ten times smaller. This would mean we would not have to consider faster-than-light travel at all, but would instead have to explain the redshift.

Up to now, we have believed that the redshift was a manifestation of the expanding universe; objects further away from us appear to move away faster and hence the speeds (or redshifts) give a distance estimate. If the quasar 3C273 is much closer than we believed, we must ignore the distance estimate based on the redshift. In that case, what causes the redshift? What do redshifts of galaxies mean now? Is the universe really expanding? Clearly, this explanation for the faster-than-light effect raises a lot more questions than it answers, so astronomers are hesitant to accept the explanation until they are absolutely forced to.

Another more realistic explanation is possible; it involves the expansion of the component parts of the double(or triple) source away from the center point, but along a line that is nearly toward us. The expanding parts are traveling almost at the speed of light. They could give the illusion of faster-than-light travel as a consequence of the known laws of physics proposed by Einstein. Imagine a part of the source of light followed by another pulse of light a year later; we would see the pulses arriving at earth at approximately the same time since the object itself was traveling at nearly the speed of light and it was almost able to keep up with the light it sent out. We would see the direction of arrival of the pulses differing on the sky (due to the motion), but the pulses would be so closely spaced in time that we would think the object must have traveled very fast between one spot and the next. However, our simple calculations of the speed involved would be wrong because we need to consider the effects of relativity on such pulses of light. Considering the relativity, we get the illusion of faster-than-light travel. If the component parts are traveling at 99.5 percent of the speed of light we could explain the observations for 3C273.

This explanation, which is far from intuitive, also leads to other problems concerning the radio components. Since one blob of matter is redshifted and the other is blueshifted, the radiation coming from the blobs should have very different intensities as a result of the relativistic speeds (near that of light). This is because the actual wavelength of emission contracts or expands as a result of the well-known Doppler effect so that a given amount of energy in an interval of wavelength is either concentrated or spread out. If we assume that the two sources in the quasar are similar in intensity, we would expect that the two components of the expanding radio source we observe would have very different intensities. Actual observations show that they are not very different from one another in any of the cases of the faster-than-light radio sources studied so far (3C273 is not alone in this category). The actual blobs hurtling apart at such great speeds must either contain different amounts of energy that differ in a way allowing the Doppler effect to change their intensities so that we see them as being nearly equal, or the picture of the two blobs traveling at 99.5 percent of the speed of light is wrong. The former is very hard to accept, but it is not an impossible situation.

It is possible to get the illusion of faster-than-light travel in yet another way. By way of illustration, consider a pair of cosmic scissors. If we close the scissors, their tips move at a certain rate but the point at which the two blades intersect "travels" much faster. If the scissors were closing at the speed of light, the intersection point would appear to travel at a speed much greater than that of light. However, although nothing is actually traveling that fast, we nevertheless see the intersection of two lines or surfaces moving. If we placed a string of light bulbs along one of the blades of the scissors and contact with the other blade caused the bulb to light up for an instant, then, as the scissors are closed, we would see a light traveling along a line that is the intersection of the blades. From a great distance we would interpret this as a single light source traveling at some speed much greater than that at which the blades of the scissors actually move. In our cosmic case we would think that something was moving faster than light while in fact it was only an illusion. Despite the apparent simplicity of such a picture, it is thought that the conditions within the quasar would have to be very special in order to produce such an effect; this is, therefore, not an explanation that is thought likely.

To sum up, we are left with three possibilities to explain the apparent faster-than-light motions in several quasars. First, it is possible that there really are elements of matter moving apart at speeds greater than that of light. That means the Einstein theory of relativity is wrong and we would have to reconsider a lot of our basic physics. Second, the distances to quasars might be incorrectly indicated by their redshifts. In that case, we would have to explain the observed redshifts in some other way. This is also not possible in the context of our present knowledge of the physics of the universe, since we would have to question all redshift measurements and hence the expanding universe model (see chapter 28). Third, it is possible to get the illusion of faster-than-light expansion if two components are moving away from and toward us, respectively. But then it is hard to explain why the components have nearly equal intensities. In addition, we would have to explain how large clouds of matter have been accelerated to within a half a percent of the speed of light in quasars, since the amount of energy required is extraordinarily large. None of these answers is fully satisfactory, and although we do not have to worry about Einstein's laws of relativity being overthrown in the near future, we do have to worry about how bits and pieces of the central parts of quasars can possibly be thrown around at such tremendous speeds. Our knowledge of this problem will continue to improve as more observations of quasars are made by the ever-vigilant "Quasar Patrol" (the name given to one group of radio astronomers that regularly measures the sizes and intensities of quasars).

CHAPTER 25
RETESTING RELATIVITY

Some experiments in physics can only be performed in the largest laboratory that exists — the universe. The tools of the experimenter become the planets, stars, and even galaxies. Such a cosmic laboratory is needed to test the theories that tell us what gravity is.

Despite being the weakest force in nature, gravity is a persistent(and sometimes painfully obvious) influence in our daily experiences. And although we utilize gravity in some special way every moment, it remains an enigmatic force in the universe. Ever since Galileo supposedly first started dropping things off the "Leaning Tower of Pisa," humans have sought to learn more about gravity.

Three centuries ago Isaac Newton formulated the basic laws of gravity that describe how objects behave within a gravitational field. Newton's laws remained unchallenged until the beginning of this century when Albert Einstein revealed that, although they were true in most cases, the laws did not account for the effects of strong gravitational fields on light and time. Einstein developed a new, refined explanation of gravity in his theory of relativity. Today astrophysicists apply relativity to everything from predicting black holes to describing the structure and behavior of the entire universe. Meanwhile, experts continue reworking and retesting Einstein's critically important theory.

According to Einstein, light bends when it passes close to a large object such as a star. The only way to test this part of the theory is to watch what happens when light actually does pass a star — such as during a total solar eclipse. When the moon blots out the sun, it is possible to detect stars very close to the sun's position. If the light from a star has been bent while passing the sun, then the star's apparent location will be different from an observation made when the sun was not in that part of the sky.

Although such experiments were performed earlier this century, the results have never been totally satisfactory since the corona (the solar atmosphere) makes accurate measurements very tricky. The experiments confirm that light bends in passing close to the sun, but they do not show whether Einstein was right or if, instead, some more recent theories correctly describe gravity.

Since Einstein's work there have been many attempts to find a better theory explaining the effects of gravity. The differences between more recent theories and Einstein's are insignificant when we consider our perception of the everyday world. But for a physicist trying to understand exactly what makes the universe tick, the differences are substantial. Experiments for measuring to an accuracy that permits detection of these small differences require the cosmos as the laboratory.

Robert Dicke of Princeton University and his colleague Brans are proponents of a revised theory of relativity. One way to compare this theory with Einstein's involves carefully observing the way Mercury orbits the sun. Einstein predicted that the perihelion of Mercury's orbit (the point at which the planet is nearest the sun) should itself move around the sun if his theory of relativity is correct. Since Mercury's orbit is so elliptical, astronomers can measure the location of the perihelion point very accurately. It shifts 43 seconds of arc per century — not large by any standard since it would take 3 million years for the perihelion shift to trace one complete swing around the sun. Einstein's theory predicts this amount of change. However, the Brans-Dicke theory calls for a smaller shift. It explains the Mercury-observed 43 second shift per century as a combination of the Brans-Dicke theory and the effect of an oblate, or flattened, sun. If the sun is not a perfect sphere, it would affect the orbit of Mercury in a way that would erroneously lead observers to believe that Einstein was correct.

By measuring the shape of the sun, astronomers can establish if it deviates from a perfect sphere and allow for that in Einstein's theory. However, the precision required for such a measurement is very difficult to achieve. Dicke spun a disk, (perfectly matched to cover the sun's image) in front of the sun and recorded the changes in light that leaked around the edge of the disk. He concluded that the sun was not a perfect sphere and that his refinement of gravitational theory was correct.

Since then, other experimenters have discovered that the brightness of the sun varies over its surface due to the presence of bright spots. These would have led to a misleading result in Dicke's experiments. In any case, the motion of Mercury's perihelion point can be explained by both theories.

The Brans-Dicke theory predicts a slight difference in the way light bends when passing the sun compared with Einstein's original work. The difference is so small that present-day optical measurements on stars are not accurate enough to settle the question of who is right. Instead, observers rely on radio measurements.

Radio signals from distant radio sources such as quasars also bend when they pass by the sun. Rather than waiting for a total solar eclipse, measurements can be made any time that the sun moves in front of a suitable radio source. A good candidate is the strong radio-emitting quasar, 3C279, which the sun occults every October. By using a radio interferometer — a combination of several radio telescopes that can accurately observe small sources — the quasar's apparent position is measured as the sun moves in front of the source. If the radio signals are bent by gravity, the source will appear to move as the sun approaches the quasar's position.

However, radio signals can also be bent when they pass through clouds of electrons, which is what the solar corona consists of. This bending (called refraction) has nothing to do with gravity. (Refraction also occurs to light when it passes through our atmosphere.) Refraction in the corona is much larger than the bending caused by gravitation, but radio astronomers have developed a way of differentiating between two types of bending. The amount of refraction in the corona varies at different wavelengths, but the gravitational bending remains the same regardless of wavelength. By making the measurements at two radio wavelengths simultaneously, astronomers sort out one effect from the other.

Recent measurements of this bending in radio signals from quasar 3C279 and several others reveal that their apparent changes in sky position are perfectly consistent with Einstein's theories. No modification, such as proposed by Dicke and others, is needed.

These experiments required years to perfect and a definitive result was only obtained recently.

Several spacecrafts have also been involved in confirming Einstein's theories. Signals from the Mars-orbiting spacecraft, Mariner 9, were carefully measured and their bending due to Mars' gravity was determined. Results agree with solar measurements on the bending of radio signals and again verify Einstein's findings.

The small slowing down of a light (or radio) signal as it goes past a massive body such as the sun can be measured in another way. It requires that the orbit of, say, Venus be accurately determined by bouncing radar signals off the planet along various points in its orbit. Astronomers would next predict what the time delay would normally be for a radar echo when Venus is at a point in its orbit on the far side of the sun, and compare that prediction to the actual delay time. The sun's gravity should slow down the sun-grazing signal, making it infinitesimally longer than expected. The newly resurfaced 305 meter diameter radio telescope in Arecibo, Puerto Rico now makes this experiment feasible.

Several other tests for gravitational theories are now being attempted. The recent discovery of a binary pulsar allows measurements that will likely support Einstein's theories with much greater accuracy. The two objects comprising the binary pulsar — probably two neutron stars — are moving about one another in elliptical orbits, and these orbits also rotate in space. They show what is called a periastron shift (the equivalent of a perihelion shift) that has already been measured. These observations do not confirm the theory of relativity — yet. However, data gathered over the next ten years should allow another exciting prediction of relativity to be tested — gravity waves.

Gravity waves — those elusive phenomena that have not yet been detected on earth — are predicted to radiate from an object whose gravitational structure changes suddenly (for example, the collapse of a star into a neutron star or a black hole). If two objects are orbiting each other at the incredible rate the binary pulsar seems to be, it is expected that radiation in the form of gravity waves should continuously radiate from the system. Although we have no way of measuring such gravity waves directly, the effect of their radiation can be detected because the binary pulsar would speed up in its orbit from year to year depending on how much energy was being released in the form of the gravity waves. In ten years, this quantity should be known and one of relativity's most important predictions should be confirmed (or disproved).

There is as yet no evidence that makes it necessary to modify Einstein's basic work, although possibly black holes do not fit into his picture. The tests so far have all confirmed his predictions. His relativity theory, coupled with new instruments and observational techniques, continues to be the cutting edge of our knowledge about the universe.

CHAPTER 26
WAS THERE A BEGINNING?

In the beginning all was radiation! Or so cosmologists now have us believe. The most widely accepted theory for the way the universe came to be involves an unusual event some 15 billion years ago, give or take a few billion years. The unusual event was a single explosion, on an unimaginable scale, during which the matter in the universe was created. From then on the universe expanded outward to its present state. One of the games cosmologists now play (a cosmologist is one who studies the universe and its origins) involves figuring out whether this expansion will go on forever or slow down and reverse itself, so that the whole lot collapses inward again. It might also be possible that these expansions and contractions go on forever. If the expansion does go on forever the universe is said to be open; if it contracts again the universe is said to be closed. Whether it is open or closed depends on how much matter it contains. If the total amount of matter is sufficiently large, the gravitational forces overwhelm the expansion forces, causing the expansion to slow down; the universe is closed. Too little matter means that gravity never wins and we live in an open universe. One of the observational problems in cosmology involves trying to find out how much matter exists in the universe.

The visible matter is all in the form of galaxies and stars that apparently do not contain enough substance to close the universe. Invisible matter may be in the form of hydrogen molecules. This is hard to observe directly. From time to time estimates of the amount of these materials are made and one year someone says the universe must be closed and the next year someone else concludes that it is open. The question is far from being resolved.

But to return to the concept of the explosive start known as the "big-bang theory," it is similar to the religious ideas that in the beginning the heaven and the earth were created. While this is a belief that is rife among many religions on this planet, it is a surprise that astronomy should come up with this model as well. There are alternatives to such a model which will be discussed later, but first let us see why this big-bang picture is currently so popular.

The basic evidence favoring a big bang concerns the observation that all distant galaxies are moving away from us and the farthest ones are moving away the fastest. This illusion is possible if all galaxies (and there are billions in our universe) are moving away from all others; then it does not matter where you are in the universe, you would always see the same thing. To illustrate this, imagine the raisin loaf analogy. When the loaf is first put in the oven the raisins are close together; as it rises (expands) while baking, the raisins move apart. No matter which raisin you are sitting on, all the other raisins are moving away from you.

In the case of galaxies the ones farther from us are moving away from us faster, which is consistent with an expanding universe. Viewed from one side of the raisin loaf the raisins nearest you are moving away from you at some speed, but those farther away will appear to be moving faster. You can see this for yourself by drawing a line and locating raisins on it and then stretching the line and measuring how far each raisin has moved in any time interval. You will find that, in the same time, a more distant raisin will have moved farther than a nearby one.

All galaxies appear to be moving apart from one another. We know, or think we know, that this is so because the light from those galaxies is systematically shifted in wavelength toward the red part of the spectrum (redshifted). This redshift is produced by the Doppler effect, which predicts that if some object emitting a wave is moving away from you, then the wave will be stretched out as seen by you. The Doppler effect produces the characteristic changes in sound of a jet plane flying by overhead. In the case of light, waves stretching out in this way means that the light becomes redder. When we examine light from distant galaxies and measure the redshift, the more distant the galaxy (based on other ways of finding distances to them, chapter 13), the greater the redshift; assuming that the redshift is due to the Doppler effect, we conclude that we live in an expanding universe.

The most recent efforts to measure redshifts show that for every million light-years away from us, an average galaxy will be moving 18 kilometers per second (km/sec) faster. If we measure the redshift of a galaxy and get its value in km/sec, then we can make an immediate estimate of its distance.

In reversing the expansion model we have to conclude that in the past, galaxies must have been closer together and, hence, at some time they must all have been in one spot. At that time the universe was unimaginably dense and hot. How long ago did this happen?

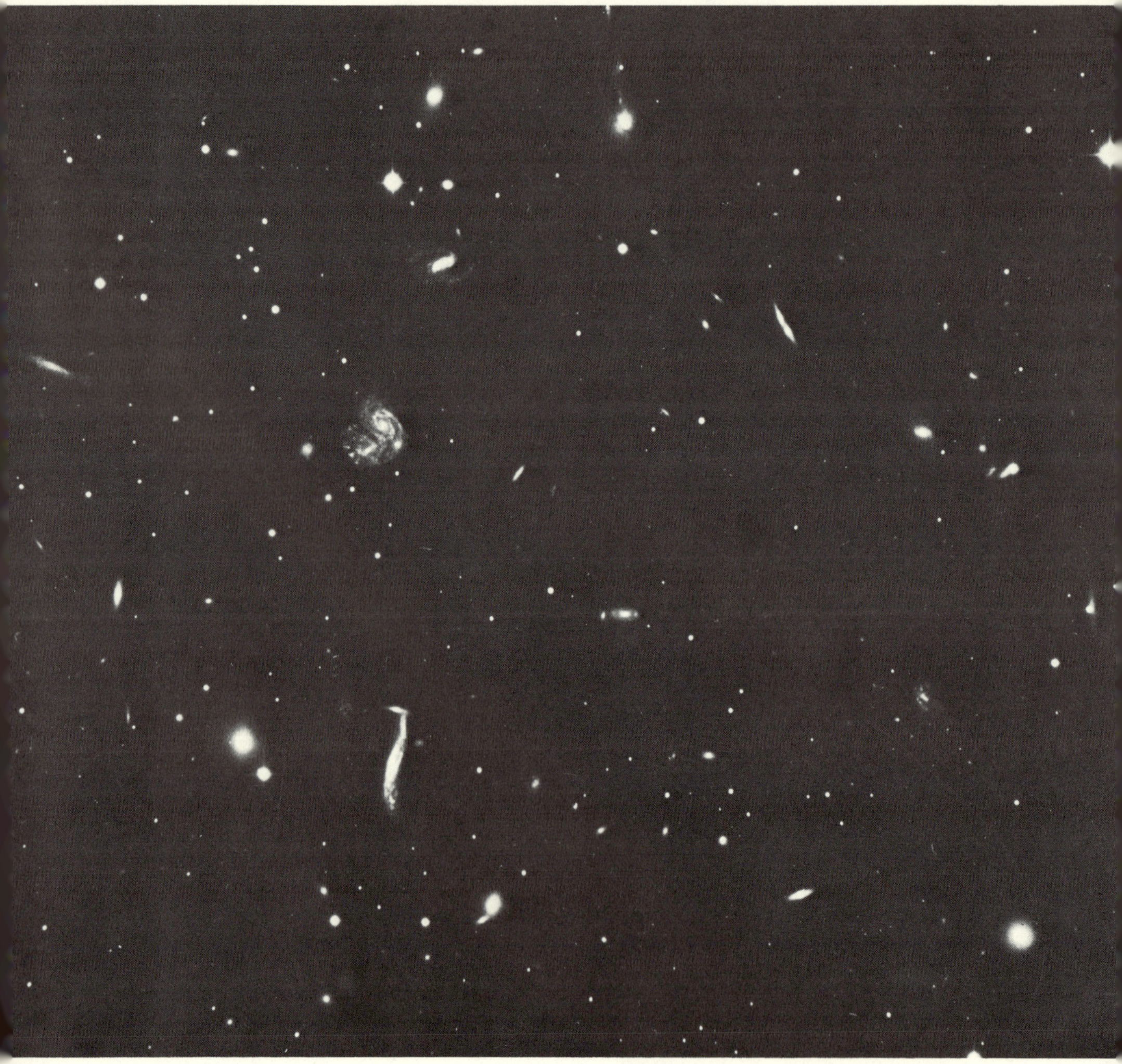

An easy way to find this is to calculate how long it would take to travel one million light-years at the speed of 18 km/sec. In other words, since an object now one million light-years away is traveling away from us at 18 km/sec, we ask how long ago both we and the object would have been at the same place. The answer is 20 billion years.

If the expansion has remained constant (and our present estimate is that the value applies to all time), then 20 billion years is the age of the universe. It is thought that the universe may be slowing its expansion somewhat, but by how much is not certain. An age of 10 to 15 billion years may be more reasonable. There are other ways for estimating the age of the universe more carefully, but we will not consider those now. Instead, let us look at what alternatives there are to a big-bang theory.

The concept that the universe has a beginning at all is anathema to many. In the 1950's an alternative model was seriously considered. In that model the universe was expanding but it was thought that the universe had always been expanding and would continue to do so. Furthermore, no matter where you are, the universe always looks the same. That implies that there is matter filling in the spaces left between the galaxies as they move apart. Matter must therefore be created in the empty regions of space. That is the essence of the "steady-state" model for the universe, and its adherents thought that the creation of matter in empty space was just as likely to occur as the creation of matter in a big-bang. Of course, neither process was understood!

The steady-state model finally died when overwhelming evidence was found in support of the big bang. That evidence was the discovery that the universe is shining everywhere as if it is at a temperature of three degrees above absolute zero. In other words, no matter in which direction you looked (in this case with radio telescopes), the signals you picked up from all around indicate that all space is filled with this three-degree radiation. The only explanation for this three-degree background was provided by the big-bang theory, which stated that the remains of the very high temperature phase just after the explosion would spread out into space and be redshifted so much that they would reach our time and space having an apparent temperature of three degrees (primarily detectable at radio wavelengths). At the time of big bang the radiation existed at very much shorter wavelengths. The steady-state theory did not explain this radiation.

The concept of a moment of creation, in which the matter that we now observe in the universe was formed, seems to be supported by observations. As one looks at the galaxies (or quasars) at great distances (that is, far into the past), they appear to be packed more closely together in space. This is predicted by the big-bang theory, whereas

The cluster of galaxies in the constellation of Hercules. Close examination of the picture reveals many strangely shaped galaxies as well as interacting galaxies, normal spirals, and elliptical galaxies.

the steady-state theory predicts that all things are equally spaced no matter where they are in space or time.

A new way to avoid the big bang has been proposed by Fred Hoyle. He suggests that there are many universes located in space and time, that our local universe is bordered by another universe, and that you should think of an infinity of such universes in coexistence. It would be wrong to try to picture our universe as one in space with a boundary beyond which there is another universe. We are talking of universes in space and time — a very hard concept to picture.

Hoyle suggested that the mass of an object in any universe depends on the location of all the other matter in that particular universe. He proposed a mass field, which is analogous to an electrical field, in the sense that the electrical force experienced by a charge in an electrical field depends on where the charge is located with respect to the source of the electrical field. If there is a mass field, then the mass of a particle depends on where it is in the universe. An electron, for example, will have a different mass at different points in space.

The boundaries between Hoyle's new universes are regions of zero mass; a particle near the boundary will have a smaller mass than the same particles far from it, and it will have zero mass at the boundary. If we look at galaxies at differing distances from us, all located within our local universe, we would be seeing the more distant ones closer to the boundary and we could get the illusion that the light from those galaxies was redshifted. This is because the masses of the particles emitting the light are different in a systematic way through space and the wavelength of the emitted light depends on the mass of the particles in the atoms. The redshift is then not an indication that the galaxy is actually moving away from us, but an indication of different particle masses at different distances from us.

The mass in other universes can be negative. But the mass field can also be negative, so that as far as we are concerned, matter everywhere still has positive mass. The important point Hoyle makes is that radiation from stars and galaxies in another universe can cross the zero-mass boundary. As it does so, it gets smeared out and will appear to us as a diffuse radiation coming to us from all directions at once and having no structure. It will in fact resemble the three-degree background radiation exactly. He therefore both avoids the big-bang expanding universe and explains the three-degree background radiation at the same time. An interesting sidelight of his calculations is that the radiation we see from our neighboring universe comes from stars that have existed for as long as 150 billion years on the other side of the zero-mass boundary. Time will tell whether his new theory gains acceptance among other cosmologists and whether future observations will support his new picture of multiple universes.

For completeness, it should be mentioned that there is another way to get a redshift in galaxies by the effect of gravity. If there is a very large blob of matter such as might exist in quasars or the centers of galaxies, the gravitational pull of this matter can stretch the departing light waves producing a gravitational redshift. It is not thought that this redshift is dominant in objects like quasars, in which some show redshifts that appear too large to be gravitational, but there might be a combination (unknown at present) of both redshifts operating on distant objects. This would seriously confuse our interpretation of the observations. Until new data show that the redshift is really an illusion and until Hoyle's theory gains wide support, we use the big-bang explanation of the origin of our universe.

SECTION VIII
LIFE IN SPACE

Speculation about extraterrestrial life is gaining more solid scientific footing. But where are the other inhabited planets, and how many of them are populated by civilizations with whom we could ever make contact? The problem may be that many of them will either be too far ahead or too far behind us, making two-way contact impossible.

CHAPTER 27
WHERE IS EVERYBODY?

Based on our knowledge of star and planetary formation, planetary systems must be very common, even the rule, not only in the Milky Way, but in other galaxies as well. Is there any way we can estimate how many of these planets exist and how many are likely to be able to sustain life as we know it? Various astronomers and life scientists have arrived at a way to estimate how many advanced civilizations there might be in the Milky Way galaxy. The estimate is uncertain since we have no idea of the likelihood of the evolution of life to the technological phase. We know that stars like the sun are common and we know that planets are likely to be common. This assumption is based not only on the discovery of planets in orbit about some of the nearer stars, but also on computer simulations of the way stars and planets form. The amazing outcome of such computer models is that for any sun-like star, one is very likely to find an earth-like planet in orbit around it. Such a planet would also be located at a distance from the parent star similar to that of earth's distance to the sun. There are, therefore, likely to be a large number of earth-like planets in the Milky Way.

Once life starts elsewhere we are not sure how likely it is that it will follow a course similar to ours. Is it common for one species to emerge, to dominate, and then become capable of changing its environment to suit itself? Is it then going to evolve to a technological society and into a communicative phase on a galactic scale? Unless the living beings become communicative on such a scale, there is no way that another civilization on another planet will ever become aware of their existence. Only after radio, television, and radio telescopes are invented can one make one's presence known over interstellar distances. We on earth have reached that stage now and have the technological capability of communicating with a twin society, with twin equipment, clear across the galaxy. It would take 200,000 years for our hello to get an answer, but there is no longer a technological barrier to this experiment.

Given that many civilizations might evolve to this stage, the most critical question to ask is how long they stay there. It is all very well to be able to communicate across the galaxy, but if civilizations destroy themselves fifty years after reaching that stage in their evolution, clearly there are not going to be many around to talk to each other. For all intents and purposes we might as well be the only ones around.

The estimates of the number of technological civilizations that exist in the galaxy, despite the uncertainties, is thought to be as great as $L/10$, where L is the lifetime of the civilization (in years) in the technological or communicative phase. We will not dwell on the derivation of this number, but at present there is fair agreement that this is a good estimate. The least known value is how long the average civilization lasts. If that number is a thousand years, then we would expect only 100 planets similar to or more advanced than ours in the galaxy. If that number is 10 million, then we would expect a million communicative civilizations.

Even with a million such civilizations in the entire Milky Way, the nearest one is not likely to be next door to us, and if these civilizations are spread randomly through the Milky Way, the nearest one is expected to be several hundred light-years away. This would imply that even a two-way chat with the nearest could require about a thousand years. That is a time span over which funding for experiments is not yet dreamed of. In any case, our civilization is far from stable enough to expect that what we regard as important today will be important tomorrow — much less a thousand years hence. The upshot of these estimates is that even if communicative civilizations last 10 million years on the average, the nearest one is located at a staggering distance of 100 light-years.

This does not mean that there are not billions of planets with life on them. There probably are, but the vast majority are likely to be so far ahead or behind us that they might as well not exist as far as ever being able to contact them. If we should ever develop interstellar travel, then exploring the nearer stars might be fruitful because they might have planets on which life has evolved but is undetectable at great distances. You should bear in mind that while life has been present on earth for 3 billion years or so, there must be planets billions of years older than ours on which life may have evolved billions of years further.

The concept of life-as-we-know-it (called LAWKI by some) is important because, even after another few million years of evolution, life on this planet is likely to be very different and not as we know it at all. Therefore, the above estimate of the number of planets with communicative civilizations presupposes that they stay in our general realm of development for all those millions of years.

It seems much more likely that evolution leads in directions away from where we are now. For example, let us assume that telepathy is a real, but undeveloped, faculty of humans and other life forms. Perhaps evolution leads to development of that faculty so that in 10,000 years we would all be telepaths. Radio, television, and talking would no longer be the modes of communication on earth. If that were common in the galaxy, there might be an enormous federation of planets inhabited by telepaths who might well be aware of our existence, but who know they have to hang around and wait for us to evolve to their level. Our estimates of how many communicative civilizations there are would then be totally wrong; our definition of communication would be wrong. Civilizations might become telepathic at some stage and there may be millions of them around already but, of course, they would be forms of life-as-we-do-not-know-it!

This is an example based on possibility, but there must be many things out there that we cannot even guess about, many being forms of life we are not familiar with. Unfortunately, we can deal in thought or practical experiments only with those life forms that are basically close to us in their stage of evolution.

The question "Where is everybody?" was apparently first expressed by Enrico Fermi. It has several answers. Those who are similar to us and with whom we are likely to communicate are around, but they are spread thinly through the Milky Way. The ones that are either not as advanced as we are or are forms of life that we cannot comprehend, let alone communicate with, may be spread thickly throughout the galaxy. And until we start to explore the galaxy in spaceships, we are unlikely to ever contact any of them.

In the meantime, what can we do to contact those whose stage of development is similar to ours? My current belief is that we have to be active, not passive, in this role. Sitting around and waiting for visitors, either on the airwaves or on our planet, is not likely to bring much success. Actively searching for faint signals from distant stars might locate intelligent life in orbit about them. Searching for the television and radio signals from those civilizations appears to be the only way to proceed at this time. We should also actively beam signals at them. Of course our signals, such as radar, television, and FM have been leaking out into space for about thirty years, and any civilization 30 light-years away would by now know of our existence, if they have radio as well. However, any who are that close to us are unlikely to be similar to us, as was previously suggested. But, if by a happy coincidence they happen to be there, we might well expect some sort of answer one day. In the meantime, there is only one thing to do: search for them. Such a search will take a long time, not only because the number of stars that need to be examined is large, but also because of the faintness of the signals that are likely to be picked up. (I refer here to searches for signals not meant for us, but the leakage of normal radio signals from other civilizations.) Alternatively, we could search for signals that are deliberately being beamed toward us. Such searches have taken place. None has been successful, nor is there likely to be success in the near future.

The major question is how long a civilization at our stage of development can last. It seems incomprehensible that we should last 10 million years in our communicative phase. After all, where were we 10,000 or a million years ago? It is much more likely that there are not too many others around with whom we can talk since we are all still evolving along different paths.

Let us look again at the question, "Where is everybody?" The answer is that they are all out there, but we have yet to do something about it. We did not discover the New World by sitting and waiting for the Indians to come to Europe! Why should we expect that we will discover other inhabited planets by waiting for them to contact us? We should actively search for planets similar to earth that are inhabited by similar life forms. The only way we can now do that is by searching for signals from those planets that are similar to the signals we radiate.

It is clear that experimental searches for alien civilizations orbiting other stars will be lengthy. Should there be a civilization like us only 50 light-years away we would still need to spend hundreds of years in our search, monitoring thousands of stars and hoping they, too, will think to look our way someday. Can you imagine anyone at our present stage trying such an experiment generation after generation (and getting funding for it)? Nevertheless, scientists are now seriously proposing that it be done.

GLOSSARY

Absolute brightness (or magnitude): a measure of how much light a star actually emits.

Amino acid: a complex molecule thought to be one of the basic building blocks of more complicated molecules such as proteins that are essential to living things.

Angstrom unit: one hundred millionth of a centimeter — used as wavelength unit for light.

Apparent brightness (or magnitude): a measure of how much light reaches us from a distant star.

Association (of stars): a very loose grouping of stars, apparently originating from one initial cloud; not quite as obvious to the eye as clusters.

Astronomical unit: the distance from the earth to the sun, about 150 million kilometers or 93 million miles.

Atom: a basic unit of matter consisting of a central part (or nucleus) made of protons and neutrons with one or more electrons in orbit around it.

Big bang: the event that set the universe expanding from its point of creation; alleged to have occurred about 15–20 billion years ago.

Black hole: a mathematical and physical concept; a region of space where gravity is so strong that even light gets sucked in. Nothing can escape once it gets caught in the gravitational pull, acting within a particular distance of the center, that depends on its total mass.

Cluster (of stars): a group of stars located close to each other in space, all born from the same cloud of dust and gas. The number of stars ranges from ten to a few hundred thousand, depending on the type of cluster.

Cluster (of galaxies): a group of galaxies that were apparently all born from the same enormous cloud of matter billions of years ago. The number of galaxies involved ranges from ten to thousands in a cluster.

Cosmic ray: a particle such as a proton, electron, or the nucleus of an atom such as helium, traveling at almost the speed of light. The heavier nuclei reach the surface of the earth as true cosmic rays. The electrons can radiate energy in the form of light, radio, etc., when spiraling about magnetic fields in distant space, but cannot penetrate the earth's atmosphere.

Declination: a parameter used (in combination with right ascension) to indicate the position of astronomical objects in the sky; analogous to latitude on earth.

Doppler effect: the change in observed wavelength produced when a radiating object is moving toward or away from the observer. Motion away from the observer produces a stretching or lengthening of the wavelength sometimes called a redshift. Motion toward the observer produces a blueshift, or shortening of the wavelength.

Dust (interstellar): the fine particles of solid material in space that absorb starlight, preventing us from seeing very far into space where dust clouds are located; a good example is the Milky Way's dark regions in Cygnus.

Dust cloud: a clearly recognizable region of space containing more dust than its surroundings.

Electron: one of the three fundamental units of matter that make up an atom; carries a negative charge of electricity.

Electromagnetic wave (or radiation): the collective name given to forms of radiation generated by electrical and magnetic energies varying in a well-defined way. Forms of such waves or radiation include radio, infrared, light, ultraviolet, X-ray, and Gamma radiation differing primarily in their wavelengths.

Emission nebula: an incandescent cloud of gas in space heated by some very hot stars inside it; emits its own light, heat, and radio energy.

Frequency: the rate at which crests in a wave pattern pass an observer; measured in cycles per second, generally called "Hertz."

Galaxy: an enormous conglomeration of stars, gas, and dust, often disk-shaped and rotating, which contains up to several hundred billion stars. The Milky Way is the name given to the galaxy in which we are located.

Gas (interstellar): the gas between the stars; mostly atoms of hydrogen but includes other atoms such as sodium and calcium as well as many molecules.

Globule: a very dense dust cloud, through which virtually no light passes.

Hydrogen: the simplest and most basic atom in the universe; pictured as consisting of a single proton at its center with one electron in orbit about it.

Interstellar: another way of saying between the stars.

Ionization: the removal (or addition) of an electron from its orbit around the nucleus of an atom.

Latitude (galactic) & longitude: the position indicator, or coordinates, used to identify a position with respect to the Milky Way.

Light-year: the distance light travels in one year; about 6 trillion miles or 9 trillion kilometers.

Magnetic field: the region of space around a magnet where magnetic forces operate.

Milky Way: the name given to the broad swath of stars and faint light that cuts across the summer skies; also used as the name for our galaxy.

Molecules: combinations of atoms in stable forms; for example, the water molecule is a combination of two hydrogen atoms and one oxygen atom.

Nebula: a diffuse object in space that shines by its own or reflected light; examples are the Great nebula in Orion and planetary nebulae.

Neutron: one of the fundamental particles, found at the nucleus of atoms, that carries no electrical charge. An electron and a proton can be forced to combine to produce a neutron.

Neutron star: a superdense object consisting mostly of neutrons; the final stage in the evolution of some large stars.

Neutral hydrogen: hydrogen in its atomic state with its electron still in lowest orbit around the nucleus; ionized hydrogen that has its electron in a higher orbit.

NGC: the acronym for the New General Catalogue of nebulae and galaxies.

Nonthermal source: a source of radiation or energy where the radiation is generated by processes other than simple motion of hot (thermal) gases.

Occultation: the passage of a planet or the moon in front of a distant star.

Parallax: the apparent shift in position of a nearby object compared with distant ones, where one's viewpoint is changed from one position to another.

Planetary nebulae: large, diffuse, spherical clouds of gas (resembling planets) given off by hot, dying stars, and heated by those stars so that they radiate their own light and radio waves.

Polarization: the property of electromagnetic radiation that gives a measure of the direction in which the wave energy is vibrating as it travels; can be linear or circular in character, or the radiation can be unpolarized if all directions of vibration are present.

Proper motion: the motion of a star through space that is "proper" to it; its accrual motion through space.

Proton: the fundamental particle with a positive charge of electricity, found in the nuclei of atoms.

Protostar: the stage in the evolution of a star just before the star actually starts to generate its own energy by nuclear processes.

Pulsar: pulsating radio sources associated with rapidly spinning neutron stars.

Quasar: (quasi-stellar radio source); energetic (and mysterious) objects at great distances from us, usually beyond the most distant galaxies we can see.

Radar astronomy: the transmission of radio signals from earth and the reception of signal echoes from the moon and the planets.

Radio astronomy: the study of radio signals from objects in space. No transmissions from earth are involved.

Radio galaxy: a galaxy that emits an inordinate amount of radio intensity.

Radio source: an astronomical object emitting radio signals.

Recombination: the converse of ionization; when an electron goes back into an orbit around the nucleus of an ionized atom.

Right ascension: a parameter used (with declination) to indicate location of an object in the sky; analogous to longitude on earth.

Seismometer: a device that measures movements of the ground (for example, earthquakes, or moonquakes).

Solar prominence: typically, an arch-shaped region on the sun's limb in which gases hang suspended; controlled by magnetic fields.

Solar (or stellar) wind: a constant stream of particles radiating from the sun (or star) and lost to space forever.

Spectral line: a minimum (absorption) or a maximum (emission) in the radiation over some very small wavelength range that is often generated by absorption or emission of radiation by atoms in a gaseous state.

Spectrum: a description of the way the intensity of emission from some object depends on wavelength.

Star: a ball of hot luminous gas emitting energy because of the fusion of elements in its interior, which also generates heavier elements.

Sunspots: dark blotches on the sun that are a little cooler than their surroundings.

Supernova: a violent explosion of a massive star at the end of its life.

Thermal radiation: radiation produced as a result of processes that depend on the motion of particles related to their temperature.

Transient (solar): a bubble-shaped region in the solar atmosphere (corona) that moves rapidly outward from the sun, sometimes associated with prominences; so far, almost exclusively observed from Skylab.

Universe: the region of space we have the ability or potential to observe with our telescopes.

Wavelength: the distance between crests in a wave pattern.